The Dream of the Factory-Made House

The MIT Press

Cambridge, Massachusetts

London, England

The Dream of the Factory-Made House

Walter Gropius and Konrad Wachsmann

Gilbert Herbert

This book was set in Univers by The MIT Press Computergraphics Department and printed and bound by Halliday Lithograph in the United States of America

Library of Congress Cataloging in Publication Data

Herbert, Gilbert.
The dream of the factory-made house.

Bibliography: p.
Includes index.
1. Prefabricated houses—United States. 2. Modular coordination (Architecture) I. Title.
NA7145.H47
1984 728.3'73 84-14414
ISBN 0-262-08140-7

Konrad Wachsmann, 1928

Walter Gropius, 1920

Contents

Part II

Politics and Technics: Years of Crisis

Part III

Dream and Reality: America in War and Peace

Acknowledgments

This study was undertaken with the assistance of research grants from the Technion Fund for the Promotion of Research and the Graham Foundation for Advanced Study in the Fine Arts. The research was facilitated by the excellent resources of German architectural literature of the 1920s and early 30s, held in the Ratner Architectural Library of the Technion, Haifa, and the extensive material in the Loeb Architectural Library of Harvard University. Material on Walter Gropius derives from the Gropius Archives in the Houghton Library, Harvard University, and the Bauhaus-archiv, Berlin, and from the collection of Gropius drawings in Harvard's Busch-Reisinger Museum. For material on Konrad Wachsmann I am grateful to Mrs. Judith Wachsmann, who made it possible for me to examine the as then unclassified Wachsmann papers at the University of Southern California and in his own home. The files on prefabrication in the Albert Farwell Bemis Foundation Records (MC 66) were made available to me by the Institute Archives and Special Collections, MIT Libraries. In interviews or in correspondence, Mrs. Ise Gropius and Messrs. G. van der Kar, Nathan Mendelsohn, Rudy Wolf, and Albert Wohlstetter shared with me some of their recollections of the General Panel venture, with which they were intimately connected. In Israel Mrs. Nadia Markowicz, Mr. Max Naumburg, Architect Dov Eitan, Dr. Walter Katz, the Tuchler family, and Mr. Ruckenstein were all helpful in their reminiscences of the Hirsch, Böhler, and Christoph and Unmack houses. I am grateful to Dr. Myra Warhaftig for her assistance with respect to the Gropius-Hirsch correspondence. Ruth Zucker, Michael Meyer-Brodnitz, David Meyer, and Elisheva Schönfeldt assisted me at various times with translations, although they bear no responsibility for the manner in which I have used them. I am grateful to Dov Cernabroda, who initially drew the copper houses to my attention, and to my former students and colleagues Michael Kwiat, Silvina Sosnovsky, Avraham Keren, and Mark Donchin, who at various periods assisted me in the search for relevant information. Drora Spitz of the photographic laboratory of the Faculty of Architecture and Town Planning was responsible for the preparation of many of the illustrations, and Marion Gold typed the manuscript in draft and final form. To all these, and many unnamed helpers, I extend my deepest thanks.

This work includes material previously published in provisional form by the Documentation Unit of Architecture, Technion, Haifa: Gilbert Herbert, *Gropius, Hirsch, and the Saga of the Copper Houses*, publication no. 1, 1980; and *The Packaged House: Dream and Reality*, Publication no. 4, 1981. To the Unit for its financial and technical support, and to Orna Betser for assistance in publication, I am most grateful.

Throughout the writing of this book I have been sustained by the encouragement and support of my wife Valerie, and my efforts have been enriched by her perceptive criticism. She and my daughter Margaret have toiled at my side in libraries and archives from Jerusalem to Cambridge, Massachusetts. In appreciation I dedicate this book to them, and to the memory of my parents, who first set me on the road to learning.

Introduction

In the United States, at the end of 1941, two architects in exile from their native Germany began to collaborate on a project for industrialized modular housing, later known as the "Packaged House." For both men, Walter Gropius, then at the height of fame, and Konrad Wachsmann, a rising star, this was the culmination of many years of experience in the field of prefabrication. Both saw in it an unparalleled opportunity to bring their long-cherished dream of a factory-made house to fruition. This, their only major collaborative effort, resulted in a system both architecturally promising (with an inherent flexibility and capacity for variability) and of great technical virtuosity. As a result of its intrinsic virtues, vigorously promoted by the persuasive advocacy of Wachsmann and the powerful leverage of Gropius' international reputation, the Packaged House received wide publicity in the professional and lay press and quickly achieved not inconsiderable fame. Funding for its development was forthcoming, modestly at first through private initiative and then on a much more substantial scale through governmental loans and guarantees. An elaborate corporate structure was set up in New York and California, whose management included many prominent men rich in technical and business experience. A large war-surplus factory was acquired and a highly sophisticated automated plant installed, with a single-shift capacity to produce 10,000 houses per year. Government approval of the prefabricated house's structure and performance capabilities ensured access to mortgage financing, and the National Housing Administration promised a letter of intent to guarantee the market of the first 8,500 houses. At the end of a decade of development, not devoid of struggle and frustration nor free of crisis but always on an ascendant curve of hope and promise, the venture seemed poised on the brink of success. But only a small number of these immaculately conceived and engineered houses was actually produced and sold, and by the 1950s the entire undertaking had collapsed "not with a bang but a whimper."[1]

That this was a grievous personal disappointment to Gropius and Wachsmann need hardly be stressed. More important, it was a grave setback to the whole movement for industrially produced housing to which both men had for so long dedicated so much thought, effort, and creative energy. Few architects were better qualified to attempt the creation of a system of factory-built houses. In Germany, in the crisis-ridden but intellectually stimulating era of the Weimar Republic, each, independently of the other, had tenaciously pursued the idea of the prefabricated dwelling. With faith and with intelligence, each in his own way had promoted the goal of industrialization. In young Wachsmann's case, the contribution to the theory and practice of prefabrication had been significant; for the more eminent Gropius, it was of monumental importance.

Now, in America, Gropius and Wachsmann represented a formidable combination of ability and experience. Their product, moreover, was conceptually sophisticated and technically impeccable, and the 1940s was a period in history perhaps more propitious than any other for a venture in industrialized housing. All the auguries were favorable, but the bold undertaking failed completely, at least in a material sense. That failure had not only the elements of personal drama, as far as the chief protagonists were concerned, but consequences of much more general significance, whether it is with the technology of industrialized building that we are concerned, or with the processes of the housing market, or, more deeply, with an understanding of the meaning of "dwelling." In this book we trace the history of the factory-made house, especially as it related to the contributions of Gropius and Wachsmann. The facts of the case are interesting in themselves. They illuminate the characters and professional careers of these two eminent architects, and on analysis, they contribute to our understanding of some of the central concerns of the architect in his confrontation with technology and social forces.

Prologue
Background to a Movement

Gentlemen emigrating to the New Settlement, Swan River, on the Western Coast of Australia, will find a great advantage in having a comfortable Dwelling that can be erected in a few hours after landing, with windows, glazed doors, and locks, bolts, and the whole painted in a good and secure manner, carefully packed and delivered at the Docks, consisting of two, three, four or more roomed Houses, made to any plan that may be proposed . . . .

John Manning, manufacturer of the Portable Colonial Cottage, advertising pamphlet, c. 1830

Your General Panel Home offers solid comfort, modern beauty, convenient livability. Here is a home far advanced in design . . . . The structural elements of your General Panel Home are made in a huge factory with modern machines and special jigs and fixtures for precision structure. . . . Your General Panel Home can be built wherever you choose, and the time required at the building site to complete it is seldom more than two or three weeks. . . .

General Panel Corporation, manufacturer of the Packaged House, advertising pamphlet, c. 1946

1

The Prefabrication of The Dwelling

The Dream of the Factory-Made House

Searching for answers to the dilemma of homeless millions in a world of potential abundance, in *A Crack in the Rear-View Mirror*, Richard Bender drew attention to a statement by Henry Ford:

The term mass production is used to describe the modern method by which great quantities of a single standardized commodity are manufactured. Mass production is not merely quantity production . . . nor is it merely machine production. Mass production is the focusing upon a manufacturing project of the principles of power, accuracy, economy, system, continuity and speed. The interpretation of these principles, through studies of operation and machine development and their coordination, is the conspicuous task of management. And the normal result is a productive organization that delivers in quantities a useful commodity of standard material, workmanship and design at a minimum cost. . . .[1]

The architectural profession has long been suffering from recurring bouts of the Henry Ford syndrome (Why can't we mass-produce houses—standard, well-designed, at low cost—in the same way Ford mass-produces cars?) encapsulated in a dream of a mechanically produced mass product which Siegfried Giedion once called the "Wohnford."[2] Early in this century, using its traditional process of design and construction, it had already acknowledged its inability to master the problems of housing the urban poor (and, latterly, of housing the middle class as well). For seventy years or more in an intensive form and for a century before that in more sporadic fashion, architects, engineers, manufacturers, industrial designers, construction firms, entrepreneurs and developers, private investors, and government agencies have been involved in an ongoing search for a cheap, durable, attractive, housing product capable of being produced by industrial processes. From the Edinburgh iron founder C. D. Young's pronouncement in the 1850s that his three-story cast-iron houses "are susceptible of being carried out in ranges to any required extent, so as to form whole streets or squares,"[3] to Buckminster Fuller's eulogy to technology in the Dymaxion house, designers have been seduced by the dream of the mass-produced house and entranced by its potentialities.

In his review of prefabricated housing systems, John Burchard wrote,

It would seem that no one could read through the list of prominent names and distinguished inventors appearing in the text without some sense of humility. Here is a galaxy of well-known names; here are the fruits of incalculable hours of thought and research by able men; here are ideas that cover in principle almost everything that a human being might conceive in the field of redesign of house structure; here is mute evidence of the expenditure of thousands, nay millions, of dollars, representing the time of many brilliant men and the labor of many others. The total cost of all the effort epitomized here may well be of the order of a billion dollars.[4]

We must remember that Burchard made this statement nearly half a century ago before markets for mass-produced housing opened up: before the defense housing boom in the United States, before the resettling of returned soldiers after World War II, before the reconstruction of war-torn Europe, before Operation Breakthrough. Burchard was writing of the 1930s, a period that we may choose to regard today as the heroic period of prefabrication, charged with enthusiasm, energy, and courage, a period of exploration and experimentation in the industrial production of dwellings. This is the period when the great masters, Le Corbusier, Gropius, Frank Lloyd Wright, found it necessary to deal with the technological imperatives and social ideology of mass housing, when each in his own manner—Wright romantically, Le Corbusier ideologically, Gropius totally and with deep commitment—explored the potentials of industrialized building. This is the period when European architects of standing in the modern movement (Martin Wagner and Ernst May, Hans Poelzig and Hans Scharoun, Josef Hoffmann, Max and Bruno Taut, Otto Bartning) engaged with enthusiasm in designing prototypes for industrial production or even total systems of prefabrication, developing them in the greatest of detail. In this crusade they were joined by Richard Neutra, Lawrence Kocher, Albert Frey, Barry Byrne, Buckminster Fuller, and many other notable architects in the United States. Their efforts, it must be conceded, were rarely brought to practical fruition; successes in the field were limited, disappointments frequent. Yet despite the lack of tangible results, the experiments continued and hopes were high.

It is difficult to understand what generated this enthusiasm for the concept of the factory-made house, what kept the dream alive. The fundamental motivation, the challenge of finding a technical means of solving the housing crisis, was the obvious spur to much thought and action in the field of prefabrication. But it is hard to conceive of this being a sufficient explanation of the passion and perseverance that marked the movement. Not even for a socially committed architect like Gropius was this enough. There were obviously other, deeper satisfactions in tackling the problems of industrialized building, which separately or in conjunction helped stoke the fires of enthusiasm.

For many architects the reward was in the creative and intellectual challenge inherent in the design itself rather than with its ultimate realization. Hence Fuller: "The Dymaxion House is still as it has been for years—a theory only. Despite pragmatic criticism it has conscientiously been kept so. While theoretical it is immediately improvable by every scientific advance. . . . The Dymaxion House rather than being a fixed solution has been naught but a statement of the problem, progressively satisfiable in the latest manner."[5] If the design process is properly conceived of as a series of stages toward the concretization of the design concept,[6] then for men such as Fuller there comes a moment when the temptation to suspend the process while still fluid is great, to remove one's creation from the constraints and limitations of real world situations, and to keep one's options open for refinement, revision, or redesign. Wachsmann, too, was later to discover that there are greater satisfactions at the drawing board than in the market place. For many of the designers involved in the search for the factory-made house, there was an idealization of the problem and its solution, an exercise highly satisfying to the creative spirit but not always productive of concrete results.

If at one level the architect is dealing with pure creations of the mind, at another he is a pragmatist, an inventor of devices, a solver of problems through the design of ingenious instrumentalities. Watch an architect at work and see the disproportionate time and effort he will sometimes devote to intricate architectural details, the intense creative joy he derives from solving technical problems. For some designers, it is true, the nuts and bolts are tedious impedimenta, but for many there is a satisfaction in the immediacy, the apprehensibility, the tangibility of solving, say, the problems of designing a panel or a jointing system. Such architects do not necessarily lack broader vision; on the contrary, even within the sweep of their creative imagination, they still see—with Mies van der Rohe—that "God is in the details." Konrad Wachsmann was such a man, drawn inevitably to prefabrication by the nobility of the technological era as he saw it and, simultaneously, by the fascination of its detailed problems. These men—Fuller, Wachsmann, Gropius—saw the totality of the whole and the significance of the part as reciprocal aspects of an integrated system.

The concept of "system" was beginning to enter into the architect's thinking at that time, although the term itself was not then in common usage. In the intellectual climate of the twenties and thirties the idea of the system was emerging to find expression in diverse and challenging directions.[7] Jan Christian Smuts published his philosophy of holism, and Alfred North Whitehead his analogous theory of organic mechanism. Ludwig von Bertalanffy laid the foundations of a general systems theory, and Arthur Tansley applied parallel ideas to the environment in his concept of the ecosystem. Walter Gropius, in *Idee und Aufbau* (Idea and Organization), gave us a prevision of his philosophy of total architecture, which was in essence a comprehensive systems theory of architecture.[8]

Prefabrication was an exemplar of the systems approach in its most direct sense, its most comprehensible form. A prefabricated system comprised a series of components organized hierarchically, with each component both a clearly defined element in its own right and a part of a more comprehensive entity. This entity, the prefabricated building, was a whole consisting of parts whose relationships to each other and the whole were governed by defined laws of combination and whose meaning derived from the whole (or end product) itself. It was a model of order and integration, based upon a clear organizing principle. As such, it had a deep appeal to many architects who were uneasily seeking a new discipline in a world when time-honored architectural standards had been abandoned and, with them, that sense of design security that comes from working within an established canonical framework. The modern movement in architecture had only contempt for historicism and eclecticism; with the rejection of the styles, those principles of permanence in design that were provided for so long by the classical orders were in disrepute. But the need for a framework of stability remained some basis for architectural consensus. Prefabrication, with its modularity and standardization, with its agreed vocabulary of parts and syntax of relationships, appeared to be an acceptable, because stylistically neutral, substitute. The interwar period in Europe (and to a lesser degree in Depression America) was an era of turbulence, uncertainty, chaos. In such a period, stimulating though it may be to artistic creativity, there often comes a deep-seated desire for stability, for predictability, for order and discipline. This was true not only in the social order and in politics but also, eventually, in art. For many architects, even some of the most revolutionary, a return to order and clarity was an inner necessity. We are talking here not of brutal extremes, of an Albert Speer turning to Fascism, or of the soulless rigidity of the *Zeilenbau*, but rather of the phenomenon of Erich Mendelsohn abandoning the exuberance of expressionism for the discipline of the Columbushaus, or of Mies van der Rohe's transformation from the sensuous poetry of the Barcelona Pavilion to the immaculate, if ice-cold, reductionism of his later work in Berlin. In this renewed search for an architecture of order, the prefabricated system provided a framework of control of such inherent logic that it appeared not as a retreat from the freedom of the modern movement but rather as its affirmation. The technological imperative sat easier on the conscience than its stylistic counterpart.

Finally, in our analysis of this dream of industrially produced housing we will find, as in all dreams, substrata of fantasy and of romanticism. The factory had a strong emotive appeal for modern architects and was regarded not only by Antonio Sant' Elia but by Mendelsohn and Gropius as an evocative symbol of the new age. It was not the futurists alone who responded positively to industry's powerful image. Henry Ford's operative principles of "power, accuracy, economy, system, and speed" found their echo not only in Filippo Tommaso Marinetti's manifestos,[9] which distorted them, but in the hearts of many a more sober architect. The humming machines, the moving belt, the turning cogs, the awe-inspiring multiplication of objects produced, all this was impressive, even fascinating, to an imaginative architect in tune with "modern times." The thought that architecture itself could be the product of such splendid mechanical installations must have been well-nigh irresistible to many an architect. The idea of the prefabrication of houses contained within it the notion, of which the architect may have been perhaps subliminally aware, of the magnification of his own professional effectiveness: a house designed by an architect was but a house, whereas a system of prefabrication could generate thousands of potential dwellings. But for the romantic this was not the dominant concern: the concept of the mass production of houses was fascinating for its own sake.

A century of development 1820s–1920s

From the present-day view those engaged in the prefabrication of houses during the heroic period, the decades of the twenties and thirties, are usually regarded as being in the very vanguard of a new movement. Gropius and his colleagues are thus acclaimed as pioneers. In a sense they are—if the movement is defined in philosophical terms deriving from the principles of mass production enunciated by Henry Ford and is directed toward the expanding market of mass housing. But with slightly different perspectives we must acknowledge that there is a prehistory of prefabrication going back to the beginning of the nineteenth century or even earlier.

During the nineteenth century prefabrication (the manufacture of buildings in component form in workshops for transport to and ultimate assembly on a remote building site) developed from modest beginnings into an industry of quite substantial proportions.[10] From the joinery shops and iron foundries, the rolling mills and shipyards, from the specialized manufacturers of Britain, the continent of Europe, the United States of America, eventually even the countries of the colonial empires, there was a considerable outflow of buildings and structures in component form. These were destined for assembly and erection, occasionally in the home market, but predominantly in an astonishing variety of export markets embracing Europe, Africa, Asia, the Americas, and Australasia. There was an impressive range of products: hospitals and schools, warehouses and factories, market buildings and stores, churches and meeting halls, barracks and blockhouses, lighthouses and bridges, theaters and exhibition pavilions, offices and arcades, conservatories and farm buildings, gasworks and railway stations. They were produced in small workshops and large industrial plants employing a thousand men, businesses styling themselves variously as producers of iron churches, portable cottages, temporary buildings, which in sum constituted a new industry known today, but only since the 1930s, as the prefabrication industry.

This industry utilized the tools and techniques of the industrial revolution, working in timber, in corrugated and cast iron, eventually in steel and reinforced concrete, using glass with growing freedom, exploiting new materials such as sheet roofing and linoleum flooring, and making excursions into such esoterica as faience and papier-mâché. For distribution which was integral to the whole system of prefabrication, there was the ever-increasing use of the new technologies of transportation, the railway train and the steamship. Its trade depended on the techniques of international commerce, developed to a high degree in the nineteenth century: an extensive banking network and intensive advertising. The buildings produced were as ambitious as the Crystal Palace, as complex as the 2,200-bed Renkioi Hospital, and as architecturally pretentious as the eclectic cast-iron facades of the New York and Glasgow loft buildings.

Right from the earliest days, among this imposing array of prefabricated products, one building type stands predominant in quantity produced and in significance. This is the factory-made house. Prefabrication in the nineteenth century must be seen primarily as an instrument of new settlement, a technological aid to facilitate the opening up of vast territories in Africa, America, or Australia to habitation, development, and economic exploitation. Whether we are talking of the founding of a colony, the drive to the West, or the frenzy of a gold strike shelter is the basic issue, housing the primary need. The imperative of immediate shelter, however modestly defined, with a modicum of security and comfort was the prime motivation for the development of early prefabrication. We know it, not only with the wisdom of hindsight but from the testimony of some of the pioneer manufacturers themselves.[11]

The sponsors of colonial settlement encouraged the use of pre-made housing, for obvious utilitarian reasons. In 1820, when Britain sent out about 5,000 settlers to the eastern Cape Province, it not only offered to supply them with good cheap tents but, on the initiative of the Colonial Office,[12] sent out several demountable three-room wooden cottages of weatherboard construction. The "Portable Colonial Cottage" produced by Manning of London in the 1830s was a much more advanced concept which necessitated a minimum of site work. It was the pioneer fully prefabricated dwelling and was an essential ingredient in the settlement of South Australia. Recent research has revealed that there were large numbers of these prefabricated houses erected in Adelaide and its environs.[13] They were also to be found right across the continent of Australia, from Perth and Fremantle in the west to New South Wales in the east and Tasmania in the south. Precut wooden houses and houses of panel construction, originating in the eastern United States, provided an answer to the urgent housing needs of the California gold rush, together with many iron buildings coming from New York, Manchester, and Liverpool. Wooden prefabs also played a significant role in the settlement and development of the American midwest, in the 1860s.[14]

Corrugated iron houses in considerable numbers were to be found not only in California but in the colonial boom towns in Victoria, the Cape Province, and the Transvaal, following the discovery of gold and diamonds. Johannesburg, for instance, established in 1886, was a veritable city of pre-made corrugated iron structures. If many of these houses were modest in size and lacking in convenience, little more than wood and iron huts, many others were commodious, well-built, comfortably outfitted, suitable accommodations for a man of affairs, a bishop, a mine manager, even a prime minister.[15] The architectural character of these prefabs was generally utilitarian and austere, but in more substantial cast iron the iron founders of Scotland produced mansions for export that combined a high degree of comfort with elegance of style and richness of decoration. So durable was the construction that these homes still provide an attractive environment for living today.[16]

At the turn of the century the attention of those interested in prefabrication was drawn to the possibilities of reinforced concrete. Hennebique's system, and the patented system of Ransome in the United States, had obvious applications in large-scale industrial buildings, but in the concept of precasting there was also a potential for exploitation by the makers of houses, who generally were interested in a smaller, and lighter, scale of components. Manufacturers were quick to develop these possibilities. By 1903, for instance, Calway and Co., manufacturers of patented cement slabs, claimed that these elements could be easily assembled by unskilled labor, to make bungalows, farmhouses, cottages, and portable buildings. It was, they asserted, "the method of the future."[17] For once, such hyperbole proved accurate.

On a more ambitious and more significant level, J. A. Brodie, the city engineer of Liverpool, devised a system of panel construction in reinforced concrete and conducted a series of experimental projects culminating, in 1905, in the erection of a three-story 12-apartment block of flats, an impressive achievement for that time.[18] Brodie's work influenced Grosvenor Atterbury, a New York architect who had made a study of it. After several years of studying the technical problems of low-cost housing, in

1908 Atterbury developed a system of construction based on hollow-cored, precast concrete units with story-height wall panels. The components were factory made and, like Brodie's, transported to the site and hoisted into position by crane. For many years Atterbury worked closely with the Russell Sage Foundation and built several hundred units for them at Forest Hills, between the years 1910 and 1918.[19] The implications of Brodie's and Atterbury's pioneering development of prefabrication in reinforced concrete were to be seen after the war in Europe's attempts to deal with the housing crisis.

Prefabrication and Mass Housing

By the turn of the century, and up to the first world war, the making of houses in factories and workshops was an established technique. A large number of firms was engaged in the business, and trade was extensive, if highly variable. Fluctuations in the market were violent, because the market itself, linked to the accidents of events—uncertain colonization, economic bonanzas, wars, and natural disasters—was sporadic and unpredictable. For most of this period prefabrication was seen as a solution to crisis conditions rather than a normal development.

One market alone escaped the attention of the hard-headed Victorian businessmen who promoted the pre-made house so determinedly. This market, surprisingly, was potentially the most promising of all, the mass market of urban housing in the growing industrial cities of Europe and America. The techniques of prefabrication were essentially those of mass production, as was clearly understood by the manufacturers themselves. Yet nowhere in this century of progress were these factory-made products directed at what might have been the most profitable and extensive market of them all, the explosive growth of the cities and the desperate need for

1.1
John Manning, portable colonial cottage, c. 1833

1.2
J. A. Brodie, Elton Flats, Liverpool, sectional axonometric, 1904–05

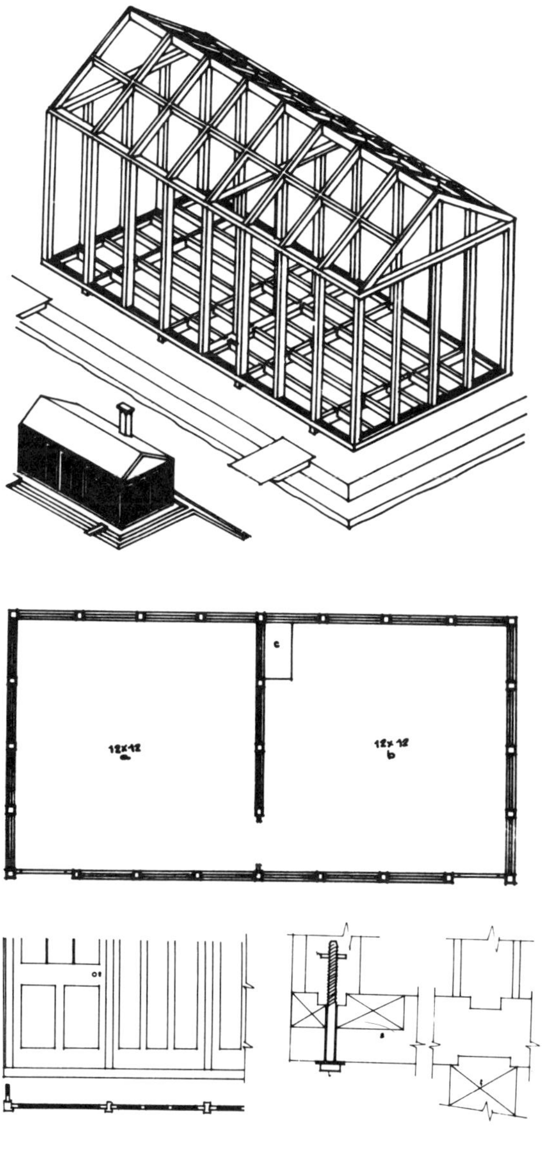

housing of the poor. Peter Thompson, a London builder and manufacturer of temporary churches had, in 1845, suggested to Edwin Chadwick that large prefabricated buildings be erected, each to house several families. But Chadwick, the housing reformer, failed to respond positively to this radical solution to the housing problem.[20] As far as we know, then, until the work of Brodie and Atterbury not one significant experiment in mass housing using industrial techniques was attempted, or even considered seriously.

This omission is particularly surprising when we consider the considerable activity in housing, much of it innovative in concept, that took place during the nineteenth century, particularly in Britain, the fountainhead of early prefabrication. Many progressive architects turned their attention to housing and, with the support of the housing associations, enlightened philanthropists, building companies, and eventually local and other governmental authorities, attempted to devise and improve individual dwelling types and the planning of housing for the working classes. Industrialists with compassion, foresight, and good business acumen conceived and built the model industrial towns (Saltaire, Port Sunlight, Bournville, Menier, Krupps, Pullman) with sensitivity to the housing needs of their workers. Yet all this activity, all this initiative and vision in architectural design, was executed on the whole in the most conservative and traditional of building techniques. The factory building at Saltaire[21] had its iron structure designed by Fairbairn, a pioneer in the manufacture of prefabricated ships and buildings; the Menier chocolate factory is a landmark in advanced iron construction. Yet what do we know of the housing of either of these two model towns, except that they were of traditional construction?

What exactly do we mean by traditional construction? The effect of the industrial revolution on building methods was felt in two ways. One of these, the revolutionary change, is expressed in what we have called the factory-built house. The other effect was evolutionary but no less important. It related not to the total building system but to the extensive use of machinery in the manufacture of traditional materials and components. Bricks were made by machine, and no longer by hand; stone was cut by power-driven saws; woodworking machinery was extensively used in the joiners' workshops; the iron foundries used the most up-to-date casting and rolling techniques for building elements; artificial portland cement replaced natural limes, and portland stone gradually usurped the place of ashlar; machine production of wire nails made them a cheap and universal product, revolutionizing timber construction; and new processes of glass manufacture removed sheet glass from the luxury list. Ready-made items from the factory—wooden mouldings, balustrades, doors and windows; iron window sashes; cast-iron columns, beams, brackets, railings, staircases; sheet floor coverings such as linoleum and felt roofing sheeting; and the ubiquitous galvanized corrugated iron for walls and roofs—became increasingly available for incorporation into conventionally built houses being erected by traditional housing firms. They were placed in position by new forms of mechanical equipment gradually being introduced on the building site. Then in the houses there was a growing component of industrially produced equipment (sanitary fittings, stoves, lighting fittings, radiators). In other words, although the building industry remained conventional in principle, there was throughout the nineteenth century and right up to the present a "creeping industrialization" of the building industry, which was often as effective in practice as it was unobstrusive to the eye. The building industry was rightly adjudged conservative, but it was not nearly as static as it was held by its critics to be.

Such improved techniques, together with the long, rich experience it had accumulated, enabled the traditional industry to achieve a predominance in the field of mass housing which was never even challenged, let alone threatened, by the new totally industrialized systems. Several factors help to explain this predominance. There was a demonstrable efficiency of the basically conservative building methods. They used tried materials and tested techniques, and still had an adequate pool of highly skilled craftsmen to call upon. They built well—unless, as jerry builders, they unscrupulously undercut traditional standards—they built quickly, and they built cheaply. Moreover, by virtue of the traditional methods and materials, they produced houses readily repaired and easily maintained by locally available skills. At the same time they were encouraged to continue with time-honored methods by the rigidity of building codes and the cautious conservatism of those authorized to enforce them: conversely, the experimental methods of the prefabricators were inhibited, at times completely frustrated, by the obduracy of authorities such as the London Metropolitan Building Office, and the inflexibility of its regulations.[22]

It may well be that the perception of the factory-made house as a temporary solution only, a perception shared by the public and the manufacturers of these buildings, inevitably prejudiced the use of these methods in the settled environment of the established cities, and mitigated against their use within the traditional urban fabric. The very terms applied by the makers to their products, the "temporary" houses, the "portable" cottages, became pejorative when considered in a more enduring architectural context. Moreover the industrial materials used in factory-made houses—wooden panels, corrugated iron sheets, cast iron plates, eventually precast concrete panels—were by the nature of the process exposed to view. This was totally unacceptable to the prevailing taste, which demanded that the use of new technology—the cupolas of the Albert Hall or the Bibliotéque Nationale, or the iron and glass roofs of the railway stations—be politely concealed from public view by eclectic architectural screens. Even in the colonies these offending materials were often later bricked round, or stuccoed over, or otherwise genteely concealed. One may readily understand that the "brutal" nature of industrial materials was offensive to nineteenth-century eyes, even those untutored by Ruskin. How much more were they inappropriate, lacking in propriety, in connection with the most sensitive issue of one's home.

The meaning of "dwelling" is the point at issue here, and it remains a critical point for prefabrication to the present day. The most conservative forces are in operation, when we build a home—and the term "conservative" is used here deliberately, with no prejudicial connotations. The function of the home is to conserve, to protect privacy, family life, cultural and social values, traditions. It is a reflection of very deep needs, for security, continuity, conformity, in an area of emotional intensity, dealing as it does with one's personal immediate environment, rich in symbolic meaning. The early prefabricated house challenged and denied most of these attributes: this was understood by the manufacturers, who thus never even presented it to their fellow citizens as an option to be rejected. In this they perhaps lacked courage, but they certainly showed sensitivity to the temper of the times.

The nature of the times changed, and our understanding of it, after World War I, when the "normality" that had obtained for a century was replaced by crisis, and crisis in postwar Europe became endemic. Therefore it is no surprise to read that "while America continued to experiment with prefabrication, Europe, by contrast, built with it."[23] The two most highly industrialized countries in Europe, Great Britain and Germany, emerged from the war with their economies in disarray, and they suffered from a housing shortage which, always chronic in the rapidly expanding cities, was now gravely aggravated by five years' cessation of routine construction. Germany, in political, social and economic turmoil, naturally took sometime to respond to the housing crisis; Britain, recovering more quickly, almost immediately began to show results in the field of industrialized building. "By 1920," according to Kelly, "the Ministry of Works had approved some 110 systems of construction, of which . . . perhaps 12 involved some degree of prefabrication." He goes on: "Between the years 1918 and 1925, a large number of partially prefabricated houses were built of elements such as sheet steel, rolled steel frames, concrete masonry, story-height precast concrete units, and expanded metal sprayed with cement. The last type of construction, combined with a steel frame, formed the basis of the Dorlonco house, some 10,000 of which were built in England between 1920 and 1928."[24]

The subsequent development of prefabrication in Britain in the 1920s has been carefully documented by R. B. White.[25] He draws attention to several significant new construction systems of the times. These include the Weir house and Atholl house, both timber framed, with external steel sheet cladding and an internal lining of millboard or fiberboard; and the Telford house, whose flanged steel plates, bolted together, formed an external load-bearing wall with felt insulation and asbestos-cement inner lining. "It is doubtful," says White, "whether more than 3,000 dwellings in Great Britain were provided in the form of steel-clad houses at this period; certainly not enough to amortize all the costs involved . . . ."[26] All these houses were relatively costly and suffered from constructive defects, especially corrosion of the exposed steel plates. They were regarded with distaste by a conservative public, and as ''makeshifts'' in an abnormal situation. ''When the shortage was overcome and the normal building methods could handle the demand, few prefabricated houses were built. They had not proved cheaper than the brick houses in whose stead they were being built, and they could not compete on even terms because of the prejudice against them as being new, untried, and substitute products.''[27] In addition the Weir house encountered fierce opposition from the building workers unions, who insisted on only their members being employed (and not the unemployed engineering workers of the shipyards) and being paid at the rates of conventional building construction. All of these factors will recur, at later stages in our story, as a leitmotif in the troubled history of the factory-made house.

Although several thousand precast concrete houses were made in Britain during the twenties, the more radical steel constructions ceased toward the end of the decade. They were not, however, without influence elsewhere. The Atholl company formed a subsidiary in Paris, and a number of houses were built in Lyons and elsewhere.[28] The Weir house proved influential in Germany where it stimulated many similar experiments. The priority given in thought and deed to mass housing in Germany, by administrators and the best known of its architects, is a salient feature of the decades following the first world war. The contribution of Gropius, May, Taut, Wagner, Alexander Klein, and others, to the refinement of the dwelling plan, the generation of new building types, the development of the technologies of mass production, and the planning of the *Siedlung* as a social as well as a physical environment, is a well-documented and fascinating chapter in the history of modern architecture. Much of this advancement

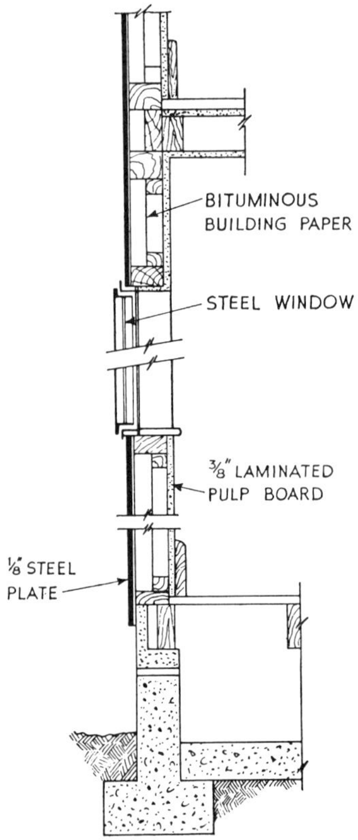

1.3
G. and J. Weir, the Weir steel house, c. 1925. *Top*: Douglas type; *bottom*: section through wall.

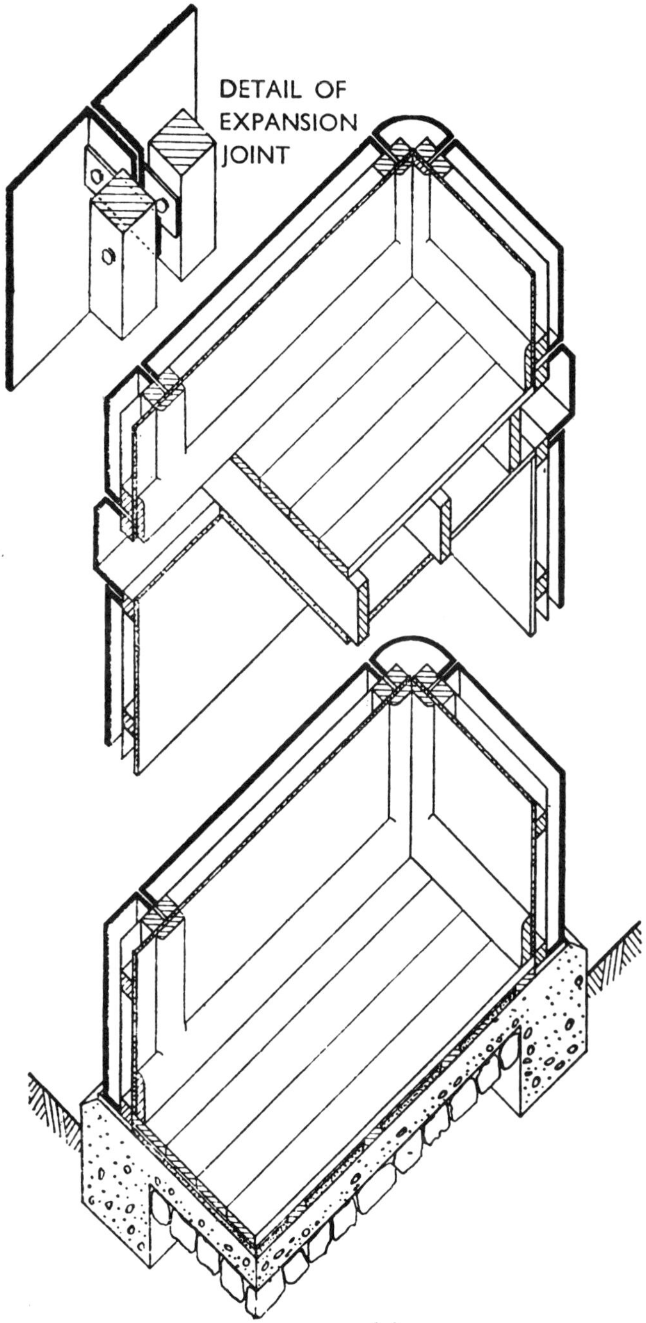

1.4
Braithwaite & Co., Telford steel house, c. 1925

was fertilized by fresh and inventive thinking, underpinned by substantial basic and applied research and development. A principal focus for this activity was the flat as the unit dwelling, to be aggregated in the new apartment blocks as the logical answer of an industrialized society to the mass-housing problem. In this context, and in an environment of creative ferment and intellectual speculation, the rationality of the flat plan was examined, a serious attempt was made to reduce it to minimal area (a major theme of the "habitation minimum" movement), and—for the first time—the question of the flexible flat plan with the provision of a fixed outer shell and adaptable inner partitioning began to receive systematic attention.[29]

Although this move to high-density mass housing was considerable and dramatic, it was by no means exclusive of other solutions. There continued, in a minor but far from insignificant key, that deep-rooted German affection for house and garden which found architectural expression in the ground-attached dwelling in its single, semidetached and row-house forms.[30] These traditional dwelling types (in both rural and urban contexts) were not only of enduring importance in the public eye and heart but evoked fresh thinking and creative design responses from the progressive architects of the day and from advanced elements of the building industry, who saw in rationalized construction methods a potent way of holding down costs. There was considerable interest in the production of low-cost houses, particularly where the price of dwelling could be fixed in advance, and then adhered to, like any industrial product.

Manufacturers of industrialized building systems cooperated with technically oriented architects to produce innovative methods of housebuilding, especially in the development of lightweight modular systems in wood and metal.[31] There was, in Weimar Germany, a ferment of ideas, in an atmosphere charged with both crisis and creativity. It is to that story of technological enterprise that we now turn our attention.

Part I

Theory and Practice: The Weimar Republic

These studies (of Gropius) in prefabrication demonstrate a theme, a clearly defined pattern. The emerging theme embraces a trilogy. It is the simultaneous satisfaction of the technical function: "to accept the challenge of the machine in all fields of production"; the aesthetic function: "the prefabrication of houses on a unified artistic basis"; and the human function: "the satisfaction of the public desire for a home with an individual appearance." The theme is the unification of technology, art and life, the fundamental unity upon which the entire structure of Gropius' concepts and works is based.

Gilbert Herbert, The Synthetic Vision of Walter Gropius, *1959*

Wachsmann's grasp of the industrialized process as one of transformation of our social and economic resources has always extended quite specifically to the transformation of our available energy media through new machine and management techniques. To harness the expanding potential of our machines and our industrialized technology, Wachsmann has long argued for a greater awareness of all aspects of the process. Unless we know the tools, we cannot possibly control them nor begin to sense the limits and possibilities out of which solutions can emerge.

Robertson Ward, ''Konrad Wachsmann,'' AIA Journal, *1972*

2

Theory and Experiment

The architect and industry

In Britain, the epicenter of the Industrial Revolution, the relationship between architect and industry in the period right up to World War I was essentially, and at its best, a guarded neutrality, which frequently degenerated into a frank and outspoken hostility. At the turn of the century, and in Edwardian days, the only native architectural tradition was that of the Arts and Crafts Movement. Ruskin was still a force to be reckoned with, William Morris still much admired. The great names in residential design—Charles Annesley Voysey, Mackay Hugh Baillie-Scott, Edwin Lutyens—were part of this solid tradition of an architecture rooted in conservative values and traditional building skills. Even when Art Nouveau was influential, as with Charles Rennie Mackintosh, it could not prize the architect free from the relentless grip of conventional construction methods, at least as far as domestic architecture was concerned. Industrial skills and techniques, those daring flights of structural ingenuity—the railway station roofs, the warehouses and dockyards—were the products not of architects but of engineers or of other, anonymous, designers. It is notable that in the century-long prehistory of prefabrication, when British manufacturers led the world in new techniques of industrialized building, architects played an insignificant role. In no instance did an architect initiate a new system of prefabrication or even play a creative part in its conception. Where the architect came into the picture, and this was the exceptional case, it was in a supportive role as the designer of a specific building in a prefabricated system already preconceived by others.[1]

It was in Germany that the first constructive links between architects and industry were assayed. Friedrich Naumann spoke at the Third German Exhibition of Arts and Crafts in Dresden in 1906 and stressed the need "to infuse mass production with meaning and spirit."[2] The following year Naumann, Karl Schmidt, of the Dresdener (later Deutsche) Werkstätten, and Hermann Muthesius were instrumental in founding the Deutscher Werkbund, when in October 1907 twelve eminent artists (many of them architects) and twelve manufacturers (involved in such applied arts as furniture making, printing, metalwork, and weaving) came to a meeting in Munich convened by Muthesius, and accepted a program whose first operative clause called for "encouragement of the fruitful cooperation of art, industry and craft."[3] It was, as it was later put, the writing of a peace treaty between art and industry.[4] This was not a purely altruistic, idealistic stand but a recognition by the founders that Germany, in order successfully to compete with the industrial giants, Britain and America, had to improve the quality of its industrial goods. The causal links in the chain—prosperity, mass production, industrialization, standardization, quality, design—were clearly recognized by the Werkbund and by most of the architects associated with it.[5]

One of these architects was Peter Behrens, who had that very month assumed his official position as artistic consultant to the AEG (the Allgemeine Electricitäts Gesellschaft) in Berlin. In this position, in which Behrens virtually became the principal industrial designer and architectural consultant to this giant firm, he had an unparalleled opportunity to realize the Werkbund's aims and his own philosophy. His viewpoint on the central issue, art and industry, was not, however, monovalent but highly complex.[6]

If he was at times moved by great engineering works, he nevertheless saw them in terms of a pseudoaesthetic, and his own predelictions in architecture were toward the monumental. He was, in a sense as Wright was, both conservative and innovative. He accepted a functional relationship between art and industrial society almost reluctantly but inescapably. The architect was compelled to work within the *Zeitgeist*, the spirit of the times; his artistic function was to transmute this spirit into an expressive force and infuse it with the eternal values of monumental architecture. He was not conservative in relation to construction techniques and went further than Muthesius in rejecting the William Morris–Arts and Crafts basis for the Werkbund: but he was conservative in his concept of the cultural role of architecture, a role he believed pure engineering could not fulfill. The challenge to Behrens, as Stanford Anderson put it, "was to bring about the synthesis of technology and art in order that modern civilization might be elevated to a true culture. . . . He wanted . . . an aesthetic 'rooted in the laws of surging life.' He also did not want a technology that pursued its own end but one that was sensible to the artistic will of the time."[7] He viewed the technological age not with enthusiasm but with an inevitability beyond debate.

One of the principal directors of AEG, Walter Rathenau, son of the founder and president of the company, Emil Rathenau, was moved by a comprehension of the times consonant with that of his architect Behrens. One could be highly critical of the mechanization of life induced by industrialization, as Rathenau was, and yet recognize industrialization and technology as obdurate facts of contemporary life, not to be wished away but to be worked with. His conclusion for Behrens or for the young Walter Gropius must have had the inescapable ring of truth: "only mechanization itself can lead us beyond mechanization."[8] Like Walter Rathenau, Gropius saw the way to a better future through the creative exploitation of industrial means. Gropius,[9] who had entered Behrens' office as an assistant late in 1907 (at the time of the founding of the Werkbund, and Behrens' new association with AEG), was one of the first architects to make a bold claim for mechanization as a way to achieve a better architecture. Like Behrens and Rathenau, he saw industrialization as a means of achieving higher cultural goals. Younger than his employer, and Rathenau their client, he was perhaps at that time a less reluctant prophet of industrial technology than they. He was burdened neither by the cultural impedance of Behrens' classicist affinities nor by Rathenau's ambivalence in the face of mechanization. This did not mean that he was unaware of the dangers inherent in the machine but that, filled with a youthful confidence that he could master it, his embrace of the potential of industry was much more wholehearted than that of his mentors.

It was to Emil Rathenau, the president of AEG, that Gropius presented a memorandum on the industrial production of buildings, in 1910, probably through the mediation of Walter Rathenau,[10] who of course stood close to both Behrens and Gropius philosophically. The importance of this document had frequently been stressed both in Gropian literature and in works on the history of prefabrication, ever since Nikolaus Pevsner drew attention to it in his 1936 edition of *Pioneers*.[11] Its contents, however, have not been subject to rigorous analysis nor have the circumstances of its preparation (even to the limited extent to which the facts are known) been probed in any serious fashion.[12] Yet such is the significance of this memorandum to our understanding of Gropius' role in the subsequent development of prefabrication that it merits the most serious consideration. Gropius himself saw it as a foundation of his work, the bedrock on which his evolving theory of prefabrication was based. It had for him a chronological significance, establishing the priority of his claims, but more than that it was, if we may use a much-abused term, a truly seminal paper.

Gropius submitted his "Programm zur Gründung einer allgemeine Hausbaugesellschaft auf künsterlich einheitlicher Grundlage, m.b.H." (Program for the Founding of a General Housing-Construction Company Following Artistically Uniform Principles) to Rathenau of AEG in April 1910, having presumably completed it the previous month.[13] It had been in preparation for some time before that. Much research, experimenting, and testing had gone into its formulation until Gropius considered it solidly based (or, as only with the presumption of youth, he could claim, it represented "the sum total of all practical, technical, and aesthetic experience"). All house types and all component parts had been designed, drawn out, detailed, specified, and costed prior to the presentation. In other words, the entire preparation had been undertaken before March 1910, perhaps even in 1909,[14] while Gropius was still in the employment of Peter Behrens, and it was presented to Rathenau of AEG, Behrens' principal client. One must assume then that it was undertaken with the knowledge and approval of Behrens.[15] The more interesting question is whether Behrens participated directly in its conception or formulation. This would appear to be a likely supposition, but there is no evidence to support it; however, as Gropius later recalled: "It is possible that I sought his advice."[16]

There can, however, be no doubt of the underlying influence of Behrens in the memorandum. Even if Gropius did not discuss the idea with Behrens, though it is difficult to see how he could fail to do so, the philosophy that guided it—the aspiration to a synthesis of art and technology in the interest of achieving a higher culture consonant with the *Zeitgeist*—is the philosophy that pervaded Behrens' office and his work. Moreover Behrens' interpretation of that philosophy in procedural and practical terms in the range of responsibilities he undertook in his work for AEG is the basis of Gropius' definition of the comprehensive role of the designer in his proposed industrialized building company.

In addition to these general considerations which constitute the ambience of the office where Gropius worked, there is another and more specific factor. At this particular time Behrens' studio was engaged in the preparation of a housing project, its first. This was the proposal to AEG for a Workers' Housing Estate at Hennigsdorf, on the Rathenaustrasse, the first section of which was erected in 1910–11.[17] Apart from some farm laborers' cottages in Pomerania which he had designed in 1906[18] prior to joining Behrens, this would have been Gropius' first serious exposure to the problem of mass housing, a problem that was to fascinate him in all the years to come. He immediately recognized it as a problem not merely of design but of a more comprehensive kind. Although the Hennigsdorf project was of conventional construction and conservative appearance, it was innovative in plan, being based on Behrens' theory of *Gruppenbauweise*, and particularly what Anderson terms Behrens' "standardized clusters."[19] The cluster plan attempted to exploit the advantages of design standardization and at the same time convey a character of uniqueness by avoiding the appearance of mechanical repetition of identical units. In so doing, Behrens introduced, though in an oblique way, the concept of standardization and variability which was to be the core of Gropius' proposal to AEG. The generation of the concept of industrialized housing was undoubtedly Gropius' significant contribution; the environment of ideas in which it germinated, its intellectual and philosophical context, was the professional office of Behrens, with its ideological links to the Werkbund, on the one hand, and its successful design intervention in the industrial world of AEG, on the other.

Gropius' proposal[20] for the industrialization of housebuilding for the mass market envisaged the production in factories of many of the components of houses, to be assembled into standard house types: workers' cottages, small and large detached houses, semidetached houses, and even flats. His proposal aimed at two objectives, which in terms of current, conventional practice appeared incompatible: improved quality of design and construction and greater economy of cost. In the formulation of these goals the dream of the factory-made house is initiated and is at the same time given its definitive form. Industrialization was the medium, as Gropius saw it, that would combine "the aesthetic activity of the architect with the economic activity of the entrepreneur," thus establishing "a happy union . . . between art and technics." The cultural significance of this development was the achievement of *künsterlische Einheitlichkeit*, an artistic unity seen as a necessary prerequisite of style. Although he used the narrow term unity, Gropius was not seeking conformity but rather consensus, which depended on a voluntary abrogation of individuality. "Conventions in the good sense of the word cannot be hoped for by emphasizing individuality. They depend on . . . the consistency of forms, recurrent because recognized as good." By these means, Gropius felt, the age could once again approach a *Zeitstil*.

In practical terms the idea would be realized by the manufacture of a wide range of housing elements (stairs, windows, doors, etc.) based on agreed standard dimensions. In order to achieve *Variabilität*, these items would be available in various qualities and materials. Through their capacity to adapt to individual needs such variants give the proposal its inherent flexibility. The client is offered a wide range of choice, within architectually predetermined limits, in house types, in mutations within each type, and in the variants in the set of factory-made elements. "It is by the provision of interchangeable parts that the Company can meet the public's desire for individuality and offer the client the pleasure of personal choice and initiative without jettisoning aesthetic unity." The concepts of user participation in the design process here articulated, and the idea of variability and choice inherent in the proposal, are innovations in architectural thinking of the greatest importance. Gropius' understanding that they are inherent in the process of industrialization anticipates that of his most advanced colleagues by a score of years: and is then rarely put forward with such clarity.

Gropius believed that the individual elements could be made by specialist subcontractors, provided they conformed to specified standards and that individual parts could be sold to outside clients. There is therefore a surprising degree of openness about the system he proposes. But this system is not comprehensive and must not be thought of as a total system of prefabrication. Nowhere in Gropius' list of elements (which includes finishes and furniture) is there any mention of major structural components: columns and beams, load-bearing panels, roof trusses; nor does he include the external enclosing elements, the walls and roof covering. For these he talks of a choice of conventional materials: brick or stone for the walling, slates or tiles for the roof. He does refer in passing to Edison's experiments in the United States with cast concrete walling,[21] but his own proposals do not go this far. Although Gropius' conception of the design implications of industrialization is so advanced, his understanding of the industrial potential is still, at this stage, quite limited, and what he proposes does not go beyond current industrial practice. In fact it falls far short of the list of pre-made items that had for years past been offered in the catalog of a building-components manufacturer such as Macfarlanes in Britain.[22]

This limited view is perhaps best expressed in Gropius' discussion of the organizational aspects of his proposed company. He sees responsibilities in three main categories: artistic, commercial, and technical. The two latter functions embrace all business aspects, with an emphasis on advertising, client contact, and sales. Nowhere is a production department mentioned. The art department, which was to be primarily responsible for the design of prototype products "developed according to basic ideas and types," was to be entirely separate from the commercial offices, although maintaining close contact with them for feedback from their contact with the public. Here the model is clearly that of the Behrens office, as consultants to, but not integral with, the AEG organization. It is interesting that whereas the public relations aspect of the proposal is detailed at length—the advertising, the informative leaflets, the articles in important journals, the lectures and exhibitions, the making of models, the display of samples—no mention is made of the designers maintaining contact with the manufacturers of components. So we have something of a paradox here: a proposal for the industrial production of houses that is conceptually advanced, highly innovative, and rich in insight but has at its core a vacuum, the complete omission of any attempt to deal with the industrial process itself. Gropius' proposal did not demand of industry anything more ambitious than could be provided by the *Handwerk* of small workshops.[23]

It was perhaps this somewhat surprising omission that contributed to AEG's failure to respond to the proposal or perhaps a lack of creative imagination on the part of the industrialists, who could not see beyond the limitations of youth and inexperience, to the real industrial potential beyond. Whatever the cause, Gropius' proposal of 1910 for industrialized houses evoked no practical response and remained in the realm of pure theory.

In the summer of 1910, just after Gropius' resignation, Le Corbusier came to work in Behrens' office. Shortly thereafter[24] he began for the first time to consider the possibilities inherent in the mass-production of houses. These thoughts first found expression in design projects: the Dom-Ino houses of 1914–15, the Monol houses of 1919, the Citrohan houses of 1921. These design exercises, mainly in reinforced concrete, show Le Corbusier extrapolating, creatively and originally, upon basic themes from Perret and Tony Garnier. His emphasis is on concrete as a modern material used to create standardized structural elements (columns, slabs) mainly through the systemization and mechanization of site procedures and on the standardization of component elements (stairs, doors, windows, cupboards) manufactured industrially. The systems proposed are always mixed systems, involving factory-made components, site-manufactured major structural elements, and a great deal of traditional handwork on the site, mainly for partitions and finishes. Far more striking than the basic technological concept is the aesthetic expression, which gives dramatic form to the idea of standardization. The second manifestation of Le Corbusier's thinking about mass production is the clarion call for industrialization of housing that appeared in *Vers une architecture* in 1923, under the heading "Maisons en série."[25] He declaims in terse, reverberating, phrases:

A great new epoch has begun.
There exists a new spirit.
Industry, overwhelming us like a flood which rolls on towards its destined end, has furnished us with new tools adapted to this new epoch, animated by the new spirit.

Then come the imperatives:

We must create the mass-production spirit.
The spirit of living in mass-construction houses.
The spirit of conceiving mass-production houses.

If we examine Le Corbusier's proposals, we see that his main points are series or mass production; standardization in both the technical and aesthetic sense, with an ongoing search for standard types; modular and dimensional coordination; and the goal of "uniformity in detail and variety in the general effect." The call for industrialization that each point subsumes had already been clearly annunciated by Gropius in his memorandum of 1910. We cannot be sure if Le Corbusier had seen this document, although it must have been a subject of office discussion when he arrived at Behrens; certainly its spirit then reflected the Werkbund ethos that permeated the Behrens studio. Conceptually Le Corbusier adds nothing new to the basic formulations of Gropius. Yet ironically in the coming years it is Le Corbusier, not Gropius, who ignites the imagination of a generation of architects; it is not the reasoned arguments of Gropius but the stimulating force of Le Corbusier's visual images, and the evocative power of his prose, that leads the Modern Movement in its drive for industrialization and standardization.

The Bauhaus Period

For more than a decade after the presentation of the abortive memorandum to Rathenau, Gropius did not return to the problem of the industrialized building process, nor did he again address the housing problem, until the 1920s. He was distracted from these earlier interests by momentous events on both a personal and national level. He entered into private practice,[26] whose main achievements, the Fagus Factory (1911) and the Werkbund model factory (1914), were buildings for industry rather than the exploitation of industry for building purposes. As the Werkbund exhibition opened in Cologne, World War I erupted, and Gropius, a serving officer in the German army, followed its cataclysmic course until the German surrender of 1918. Then, amid the economic and political chaos of the postwar era, Gropius became deeply involved in the politics of the emerging Modern Movement in architecture; he immediately plunged into what was to be his greatest challenge and his most enduring achievement, the establishment of the Bauhaus at Weimar in 1918.[27]

At Weimar, and subsequently at Dessau, the Bauhaus, while under Gropius' direction, did not deal substantially with the problems of prefabrication. Indeed it hardly dealt with any architectural problems, other than tangentially through the agency of Gropius' professional office. The isolated experiments in prefabrication under the Bauhaus aegis we will discuss later; they are interesting in themselves but do not add up to a coherent or consistent program. However, what the Bauhaus did concern itself with as basic were two issues central to the concept of industrialization generally. These issues were the relationship between art and industry and standardization of dimensional and typological norms. In other words, the Bauhaus continued to address itself to the two main themes of the Deutscher Werkbund.

Although Gropius' slogan, "Art and technology, a new unity,"[28] was not uncontested, even within the Bauhaus,[29] the evolution in education and production from a handicrafts orientation to a fuller involvement in industry took place inexorably and reached its peak with the move from Weimar to Dessau. A marketing organization was established "in order to help establish contact between industry and the Bauhaus . . . [whose] function is to take care of the sale of prototypes to those branches of industry which can mass-produce from completed prototypes and market the product."[30] The designer's function, as Gropius came to see it, was as the designer of prototypes for industrial production. He had already spelled out his hopes in this direction in 1922[31] Now, in a policy statement in 1926, he said: "The Bauhaus wants to serve in the development of present-day housing, from the simplest household appliances to the finished dwelling. . . . The home and its furnishings are mass consumer goods, and their design is more a matter of reason than a matter of passion." The machine, Gropius went on, can provide the individual "with mass-produced products that are cheaper and better than those manufactured by hand."[32] It is evident that this policy statement is still informed by the spirit of the 1910 memorandum. The Bauhaus' theoretical and practical involvement in industrial mass production, although limited to domestic consumer products (textiles, furniture, light fittings, wall paper, heating stoves) was a significant preparation for a return to a concern with the wider issue of industrialized housing. It was an experiment in the sense of a limited learning experience of directing design to the specific tools and processes of industrial production. Out of this proving ground of trial and error there were lessons of a more universal nature to be learned.

In his statement on the principles of Bauhaus production, of 1926,[33] Gropius affirmed his belief that "the creation of standard types for all practical commodities of everyday use is a social necessity." Standardization was essential, he maintained, in order to exploit the effectiveness of the machine as a device for mass production of products "that are cheaper and better than those manufactured by hand." The drive toward standardization in the Bauhaus is expressed best by the attempt to crystallize a few, ideal, solutions to everyday problems. To this end the workshops of the Bauhaus were to be regarded, in Gropius' phrase, as "laboratories in which prototypes of products suitable for mass production and typical of our time are carefully developed and constantly improved."

There is no suggestion, at this stage, of regarding these industrially produced elements as part of a comprehensive system. Within one subsystem, as in the case of unit furniture designed by Breuer,[34] the interface between elements and the overall ordering principle of a modular grid governing standard sizes and variants might be seriously studied. This study of unit furniture followed on earlier experiments to standardize furniture notably by the Deutsche Werkstätten of Karl Schmidt, who had produced *Typenmöbel* as early as 1910.[35] In the Bauhaus, at this stage, this principle of integration and order was not extended in any systematic study to the overall system of dwelling and contents. This is best demonstrated by the experimental house "Am Horn," designed by Georg Muche of the Bauhaus teaching staff and erected in Weimar as part of a projected, but never realized, Bauhaus Housing Settlement in 1923. In this house, for whose construction Gropius' office and Adolf Meyer were responsible, "industrially prefabricated products and new building materials were used *wherever possible*."[36] Even Gropius and Meyer's proposal of the same year for Baukasten im Grossen, an adaptation of the concept of children's building blocks in terms of large-scale prefabricated elements, was an experiment that examined one particular problem, that of variability within a standardized system[37] rather than the total problem of an overall system of prefabrication.

The architectural expression of this building system is, by the nature of things, cubic, boxlike, additive, austere. It creates an architectural language very different from the romanticism of Gropius' earlier postwar houses—the Sommerfeld house of 1920–21, the Otte House of 1921–22, and the Kallenbach project of 1921.[38] It forms the aesthetic basis moreover for Gropius' future projects, from the Auerbach house, in Jena, in 1923 to the more freely disposed Masters' houses at the Bauhaus at Dessau in 1925–26. The Baukasten im Grossen project then may be considered not only important for its role in the history of prefabrication but also for its links—as an expression, if not a cause—with the subsequent evolution of a modern architectural style.

The first attempt to translate industrialized building from a theoretical postulate to a practical exercise came with Gropius' Reichsheimstattensiedlung, a housing scheme at Toerten-Dessau built over the years 1926 to 1928.[39] The project was commissioned by the City of Dessau, and was based on extensive research both prior to and during the project, and sponsored after 1927 by the *Reichsforschungsgesellschaft für Wirtschaftlichkeit in Bau-und Wohnungswesen*, the national society for research into economic building and housing of which Gropius was an executive officer.[40] Planned for a population of 5,000, Toerten-Dessau was carried out in stages over three years, 1926 to 1928, under Gropius' direction, and 316 of his two-story row-house units were constructed. These were of standard design, but with modifications from year to year, and were constructed of reinforced concrete and cinder blocks. Cross walls, beams, infill blocks, floors, and roofs were standardized and were manufactured on the site. Sand and gravel found on the site were suitable for concrete, and it was only necessary to transport cement and cinder. Materials were stored on site and moved by trolley along prelaid tracks to casting areas between the houses, where the building elements were cast and cured, close to the point of use. When ready, they were hoisted by mechanical equipment and set in place by special teams. "The principle of work at the site was to reuse the same man for the same phase of the construction in each block of houses and thereby increase output," explained Gropius. "In order to insure the interlocking of the individual construction phases from the start of the rough construction and interior work, an accurate timetable was worked out, similar to the ones used by railroads."[41] The stages of construction were carefully articulated; the basic shell, for instance, was defined in fourteen constructional steps. A time chart was drawn out, and it provided both a visual survey of planned progress and a means of control. A photographic check on the progressive stages was maintained.

This project cannot be considered as prefabrication, nor did Gropius so consider it.[42] All work was carried out on the site, much of it by traditional means. Yet it is a form of industrialized building, with the organization of site operations as a whole work process analogous to the factory. We have here the concepts of standardization, mass production, specialization of labor, mechanization of operations, and rigorously planned organization of labor and materials, which are the characteristic features of the industrial system. Here these characteristics are transferred in a limited way, but with consistency, to the building process. Gropius is now exploring an alternative path to the industrialization of house construction to that adumbrated in his 1910 memorandum. In so doing, he is part of a wider movement; and Toerten-Dessau must be seen as one of several significant experiments then being undertaken.

The first of these was Martin Wagner and Bruno Taut's Hufheisensiedlung, in Berlin-Britz, in 1925–27. As early as 1920 Taut had put forward, in *Die Auflösung der Städte*,[43] a system of house construction leading both to standardization and variability. In 1924 Taut wrote: "The

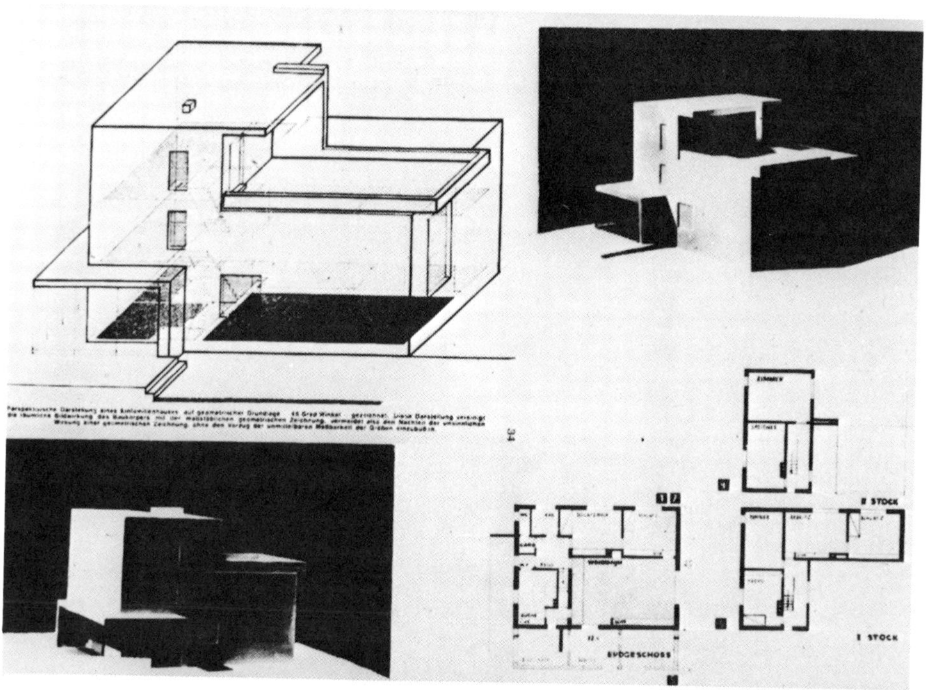

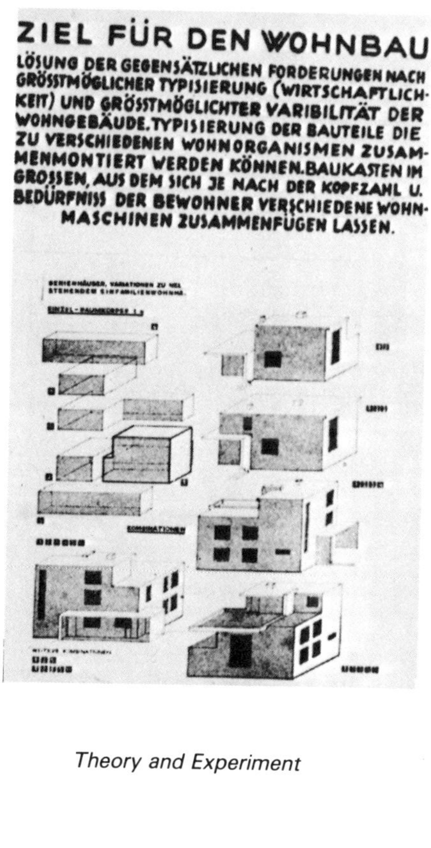

2.1
Gropius and Meyer, standardized housing using "building blocks," 1923

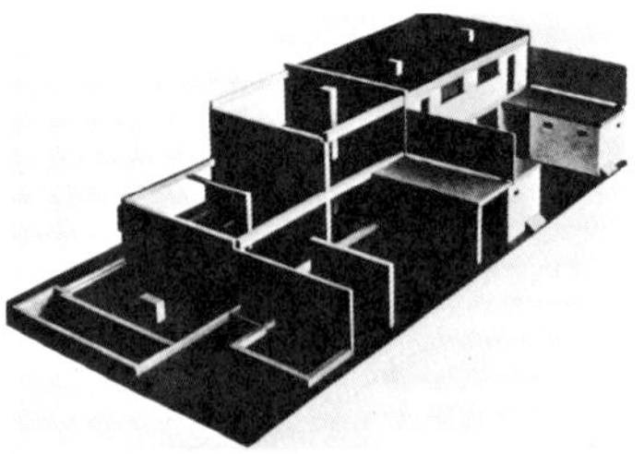

2.2
Walter Gropius, housing estate, Toerten-Dessau, construction system, 1926–28

2.3
Walter Gropius, housing estate, Toerten-Dessau, site organization diagram, 1926

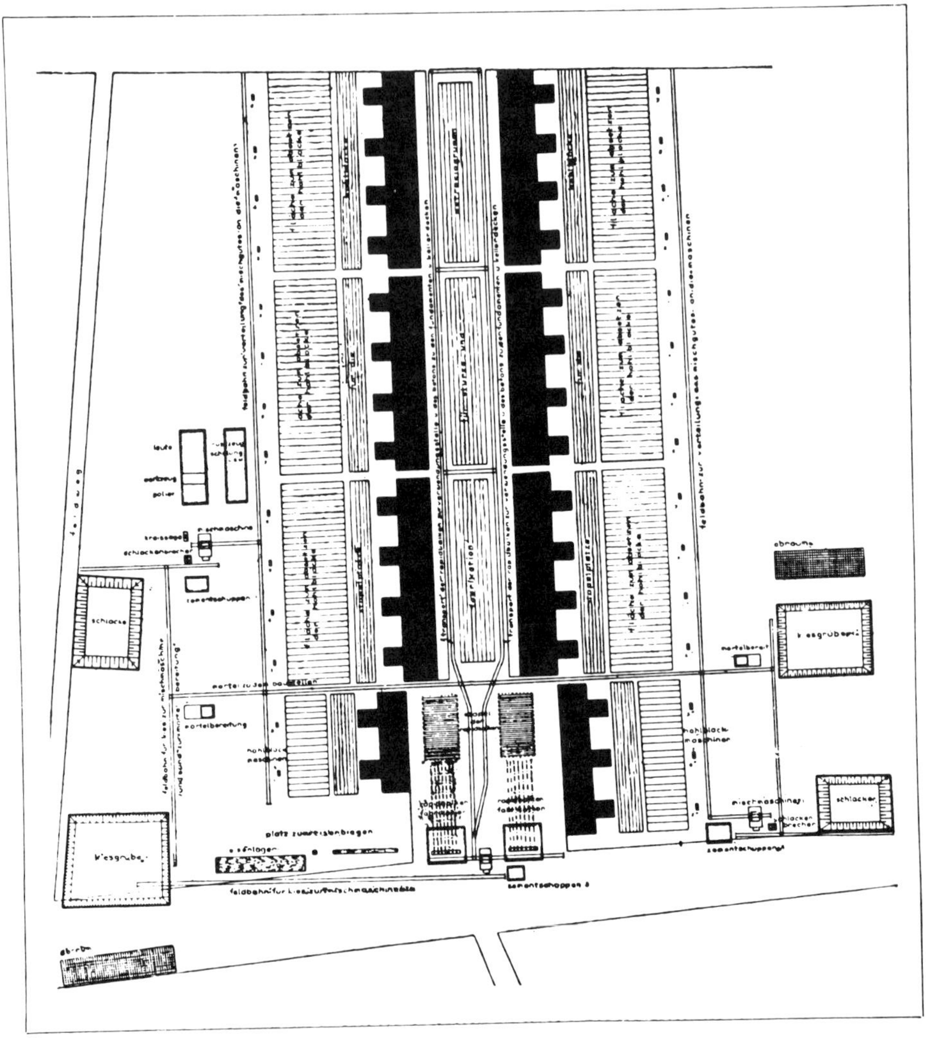

2.4
Bron System, housing estate, Berlin, c. 1926

problem of house-building today must be tackled along lines that are valid in industry for the production of machines, cars and similar objects."[44] Now, in 1925, he approached this concept in practice, together with Wagner, tentatively and from a somewhat different point of view, as Gropius was soon to do at Dessau. Utilizing the cooperative building construction company organized by Martin Wagner, the two architects planned the comprehensive housing estate at Berlin-Britz in which they made use of large-scale mechanical equipment on the site such as traveling cranes, then an innovation, and set up a rigorous division of labor that involved specialization for specific repetitive tasks.[45]

A much more advanced scheme, built at about the same time as the first stage at Toerten-Dessau, was initiated in Friedrichsfelde, Berlin, early in 1926.[46] Here a group of 31 three-story buildings was erected, using a precast concrete system of construction based on the Bron patent, a Dutch method used for the first time in Germany. Precast concrete construction was of course experimented with not only in Germany at this time but elsewhere in Europe. In addition to this Dutch system we also have notable examples in France, which, since Hennebique and Perret Frères, had been a center of experiment and development.[47] The Bron system involved the casting of large story-height wall panels, complete with their windows and doors and all other components such as beams and slabs on the site, and then transporting them by a large overhead crane moving on tracks that straddled the line of buildings under construction.

This approach was developed even further in a large-scale housing scheme at Frankfurt, where at Praunheim in 1926–30 Ernst May built 1,400 dwellings.[48] Again rationalization and mass production were the key principles. Plans were limited to a few carefully thought-out types, and many details were standardized. *Frankfurter Normen*, whose use was obligatory if a mortgage was desired, were laid down for doors, hardware, stoves, sanitary ware; the highly efficient "Frankfurt kitchen" was designed for standard use; and a construction system using prefabricated large-scale universal concrete panels, all 3.0 by 1.1 m, and precast beams, was developed. This system was produced by the Frankfurter Montageverfahren, set up by May in a large empty machine hall, with the capacity to turn out a standard slab in three to five minutes, and was further promoted by the Reichsforschungsgesellschaft, the research organization in which, as we have seen, both May and Gropius were leading figures.

In all this activity we find the production emphasis in industrialization being shifted from the factory to the site, the technical solution becoming specific in relation to each project and no longer universal, and the conceptual solution to the housing problem moving inevitably from the private house (the main thrust of Gropius' memorandum to AEG in 1910) to mass housing, to the row houses and apartment blocks of the *Siedlungen*. The reasons for this expansion of scale may be found in the social and economic conditions of the times; in the magnitude of the housing shortage;[49] in the recovering economic situation of the postinflationary era and the considerable increase in housing investment; and in the highly volatile social and political situation, potentially threatening, for which housing was seen as an anodyne.

By the end of the decade, however, the pendulum had once more swung back. The promising economy of the late twenties became a casualty to world depression. Unemployment soared, and private investment in housing declined. Public spending on housing was cut drastically with the halving of the special rent tax, and new construction, as a consequence, was drastically curtailed.[50]

2.5
Ernst May, the Frankfurt building panel factory system

The government, alarmed at the long-term prospects, began to encourage a return from town to country and the building of small cottages for workers. The back-to-the-land movement added an economic dimension to its inherent romantic appeal. After a few years of spectacular success the age of the *Grosssiedlung* was suddenly at an end and, with it, the brave program of experiment in the industrialization of mass housing, a program that, at Berlin, Frankfurt, and Dessau, was still in its infancy.

In the very heartland of the Modern Movement, with the most stalwart supporters of large-scale mass housing, interest in the one-family house revived, if indeed it had ever been abandoned. Perhaps it is nearer to the truth to say that work on the small house proceeded on a track parallel to that of the *Siedlung*. The design of the one-family house had always demanded a share of the architect's creative ability and interest disproportionate to its size but, psychologically speaking, commensurate with its significance as an environmental problem. The one-to-one relationship of family to house was close to the architect's heart, and even the advocates of apartment buildings often agreed with May's ideology that the ground-attached individual dwelling was the preferred solution to the problem of living. It was to be regretted therefore that it was not economically attainable, unless—and here the dream of the factory-made house reasserts itself—machine production could significantly reduce costs.

It is entirely understandable therefore that the problem of the prefabricated house continued to engage the inquiring spirit of Gropius and his colleagues at the Bauhaus, even at the time when they were involved in the large-scale Toerten-Dessau housing scheme. The concept of the house as an industrial product was consistent with the Bauhaus philosophy; it epitomized at the most significant level, that of architecture itself, the Bauhaus vision of the unity of art and technology. It was fitting therefore that in December 1926, to coincide with the opening of the new Bauhaus building in Dessau, the Bauhaus was instrumental in erecting on a site at Toerten an industrially produced house made of steel. Georg Muche, the form-master in the weaving workshop of the Bauhaus whose experimental house "Am Horn" in Weimar we have already noted, was reponsible for this bold gesture, together with the architect Richard Paulick.[51]

The Muche-Paulick steel house was manufactured by Carl Kästner and Co. of Leipzig, whose experience lay in the making of special machinery and in the production of safes and strong rooms. When Kästner began to experiment with utilizing their technical expertise and metal-working tools for the manufacture of steel houses, they turned to a system of wall construction recently developed by the firm of Braune and Roth, who had also been exploring the possibilities of steel house manufacture.[52] Some of Braune and Roth's experimental products, ranging from weekend cottages to a large, six-roomed two-story house, were exhibited in Berlin and Leipzig, with external cladding that consisted of large, 2 by 3m steel sheets.[53] The Kästner system used a similar construction, with a light framing of steel I-section (or double T-section) posts. This frame was clad externally with 3-mm-thick Siemens-Martins steel plates, sealed with a T-shaped cover strip complete with a rubberized gasket at the joints. A 60-mm air space separated this external skin from the inner lining of "torfoleum" (wood-wool on peat fiber base) insulation boards and 50-mm-thick cinder concrete slabs which were plastered internally.

A demonstration house erected by the Kästner company in Leipzig shows a strange union of a radical wall structure whose steel panels are unambiguously exposed, and a conventional steep pitched roof complete with dormer

2.6
Muche and Paulick, steel house, Dessau, 1926. *Top*: view, as erected early 1927; *bottom*: plans.

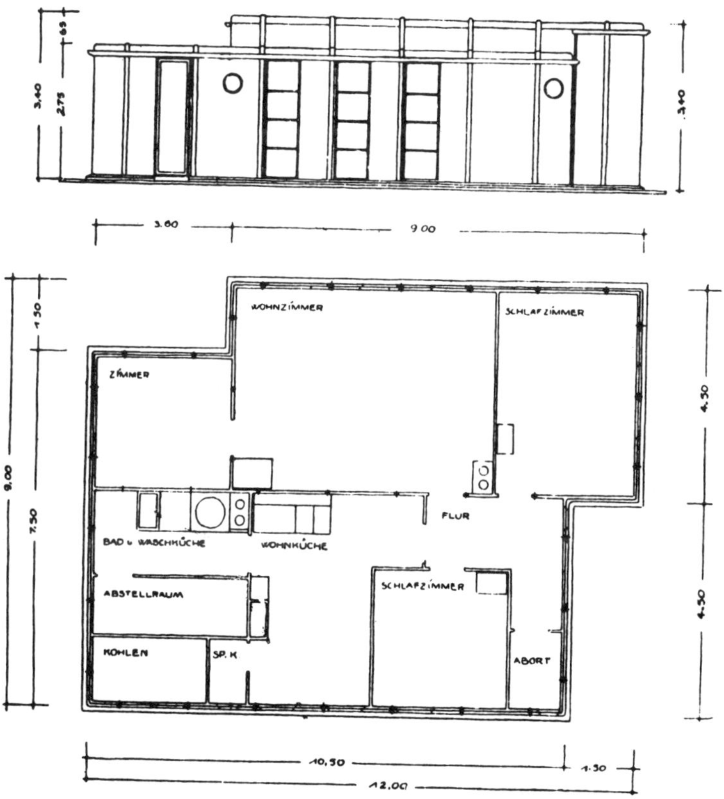

window. The Bauhaus model uses a similar constructive principle but is completely different in appearance. It has a flat roof of steel beams and concrete slabs, duly insulated, and a tighter silhouette, although it is a composition of two prisms of different heights, a carry-over from the Weimar Baukasten project. The modular principle rigidly adhered to in the structure does not appear to generate a planning grid, however, and internal partitions do not lie on the modular lines. Inevitably, the 1.5-m module is too wide for the doors, and the windows which are narrow vertical strips, fill only part of a module. Nevertheless, Muche claimed for the house an inherent variability of plan arrangement, with the possibility of moving internal partitions. He foresaw that with mass production the repetitive use of standard profiles and component elements would inevitably result in an economic product.

In that the house involved industrially produced main elements, including the frame, the cladding, the partitions, and the roof, it went far beyond the more limited scope of the 1910 memorandum. But the credit for this should go, not to Muche and Paulick, but rather to Braune and Roth whose system formed the basis of the Bauhaus-designed house. And we should note that Braune and Roth were themselves not operating in isolation but within the context of an emerging German prefabrication industry much concerned with steel houses. To that important topic, however, we must return in a later chapter. For Muche, this was a one-time study, a fascinating design challenge, and he made no further attempt to develop the prototype for industrial production. For Gropius, the problem of the factory-made house had much deeper roots and much wider implications. But the next stage of his investigations was also limited to a case-study house, with no immediate practical consequences. This house was the widely publicized prefabricated house at the

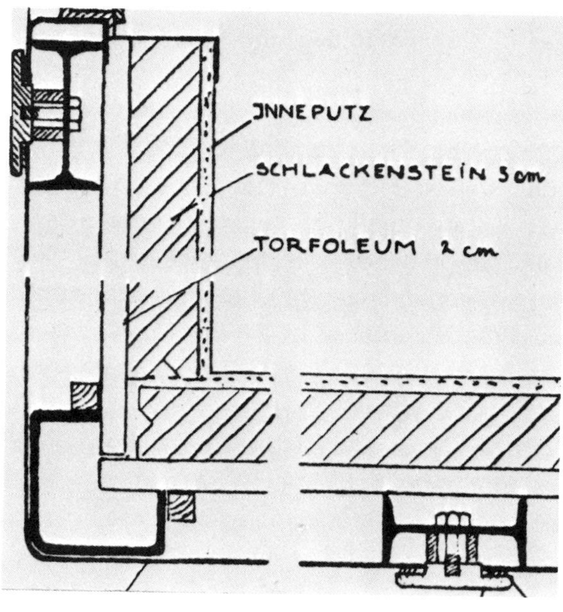

2.7
Carl Kästner Co., steel house, Leipzig, c. 1926

Weissenhofsiedlung, the Deutscher Werkbund model neighborhood of 1926–29, in Stuttgart.[54]

This house was one of two houses designed by Gropius for the exhibition, the other being only partially prefabricated. The fully fabricated house is a two-story structure, whose whole plan is reduced to a simple rectangle and whose form to an elementary prism. The term Gropius uses to describe its construction system is *Trockenmontage*, a dry assembly system. Upon an in-situ concrete foundation (the only exception to the "dry" rule, and a persistent problem in prefabrication), a steel frame is set up, consisting of Z-section uprights, channel section horizontals, and I-beams for floors. This frame is clad with asbestos sheeting on the outside, lignat sheeting (a cellulose fiber product) internally, and there are 80-mm-thick pressed cork slabs in between, as insulation, separated from each of the wall linings by an air space of 30 mm. The roof is of precast cinder concrete blocks covered with metal. Floors are of wood, and ceilings of celotex sheets. Although the system resembles the Muche-Paulick system in some respects—the steel frame, the external sheet cladding, the insulation—it differs in its inner construction, which avoids such "wet" trades and in-situ work as plastering. The structural module is 1.06 m, to accommodate a standard door frame. The planning module follows the structural grid, and all internal partitions are located on grid lines. It is apparent that Gropius has far greater understanding here of the design discipline that stems from the acceptance of a modular principle of structure.

The system remains elementary when broken down into its constituent components. Each steel section, wall lining, floor board, and insulation slab is a separate industrially produced item. All have to be assembled on the site.

There is no concept yet of subassemblies of entire building elements, such as wall panels or inner partitions, arriving ready-made from the factory. In other words, the system remains a primary one, with no clearly defined subsystems. It is essentially an open system, not dependent on the integrated output of a single production belt but incorporating a large variety of established industrial products of diverse origins. As a result the system is a flexible one, providing many alternatives, and responsive to change. The price paid is a considerable amount of site work, albeit of an assembly nature rather than construction in its traditional sense. But within these limitations we do have what was lacking in Gropius' memorandum of 1910, a coherent overall system. It is moreover the first system of prefabrication conceived and designed in all its technical details, if not by any architect, then certainly by an architect of Gropius' professional stature. The fact that this work was given great public and professional exposure, directly and in the press, because of its inclusion in the exhibition of the Weissenhofsiedlung, was of great importance to the movement for industrially built houses. Gropius therefore provided this movement with one if its first convincing practical demonstrations. He was to continue to be deeply involved in these empiric endeavours, as we shall see. But at this time his more important role was as the formulator and chief propagator of a coherent theory of industrialized building.

The Theory of the Factory-Made House

The erection of the experimental houses at the Weissenhofsiedlung was an opportunity for Gropius once again to formulate his thoughts on the general principles of prefabrication as a means of lowering housing costs and improving housing quality—the goals, it will be recalled, of his 1910 memorandum. These general principles he dis-

2.8
Walter Gropius, prefabricated house, Weissenhofsiedlung, Stuttgart, 1927. *Left*: view; *right*: construction details.

cussed in a short but important paper: "Wie bauen wir billigere, bessere, schönere Wohnungen?" (How Do We Build Cheaper, Better, More Attractive Dwellings?)[55] We may summarize the essentials for achieving his ultimate objective, "das fix und fertig eingerichtete variable Wohnhaus" (the ready-furnished [that is, fully equipped and ready-to-use] adaptable dwelling) in the following points. Houses should be made by mass-production methods, using the assembly-line production processes of factories, and monitored by precise flowcharts and other control methods. The elements of the houses should be made for dry-assembly construction systems, utilizing the newest building materials. The houses should be designed according to rationalized building plans and should be carefully studied to the very last detail in a way comparable to the meticulous design of machines. Such a design process postulates the idea of standardization of the product; this product, however, should not be the whole house but the various constituent components comprising the building system. Such a stock of standard elements would enable the erection of houses that could be varied according to the customer's demand (within of course the limits of the system). It should be possible to fix the cost in advance and to guarantee the construction time. In such a rational production system, independent of the variable of the weather, long-range financial planning becomes possible, and the funding of research is of major importance. Finally, in this paper, Gropius proposes that the direction of such a project should be in the hands of a team comprising architect, engineer, and businessman, each with his clearly defined role to play.

This is by far the most comprehensive statement on the principles of industrialization of houses to date. It is compatible in its aims and objectives with the original memorandum to Rathenau but is much wider in scope, dealing as it does with total rather than partial systems and showing a deeper understanding of the industrial process. The guiding philosophy underlying these principles was once again stressed by Gropius in a paper written shortly thereafter, and published in the second issue of the *Bauhaus* journal in 1927: "Systematische Vorarbeit für rationellen Wohnungsbau" (Systematic Preparation for Rationalized Housing Construction).[56] Here Gropius states: "The ultimate objective of this trend will be accomplished only when all the reasonable wishes of the individual for his home can be fulfilled without sacrificing the economic advantages of mass production. The houses and their furnishings will differ in their general appearance to suit the number and kind of their inhabitants. On the other hand, the components from which these buildings will be made will be identical. The 'type' itself is no obstacle to cultural development; on the contrary, it is almost one of its prerequisites. . . ." Gropius, in order to achieve the goals inherent in his theory of prefabrication, had to set the process in a wider context. He therefore calls for a rationalization of the entire building industry, to be studied within the framework of a system of national and regional building research institutes. The program of work of these institutes, embracing both theoretical and practical aspects, constitutes the "systematic preparation for rationalized housing construction" which is the main argument of his paper. The following are the main points of his interest: the optimization of dwelling design, based on generalized studies of user requirements; the maximum exploitation of new materials and techniques of construction, and specifically techniques of prefabrication; the stockpiling of pre-made components, manufactured according to agreed norms and standards; the rationalization of national distribution systems, especially highway development; the study of the efficiency of building site procedures, with respect to materials, labor, and mechanized equipment; the institution of comparative costing and estimating procedures; and a revision of building legislation to permit the exploitation of new technological developments. The range of Gropius' thinking here is impressive, as he sees the problem in many-faceted form; also impressive is his humanism, as his search for norms and standards is always explicitly based not only on technical and economic considerations but on what is socially desirable. Prefabrication, or mass production through industrial means as Gropius sees it, is not only a search for quantity but essentially a means of achieving quality.

It is this search for quality that brings Gropius to the other element in his theory of industrialized housing—the central role of the architect. It is the essence of the third major theoretical paper which we have chosen to consider briefly here: "Der Architekt als Organisator der modernen Bauwirtschaft und seine Forderungen an die Industrie" (The Architect as Organizer of the Modern Building Industry and His Demands on Industry).[57] Here we have a long philosophical discourse on the schism between art and industry since the days of the medieval craftsman, who embraced within his personal skills the functions of artist, technician, and tradesman. This was no longer possible within the framework of an industrial society, where the opposition of technical and formal considerations was inherent. Who is to resolve these conflicts? Who is to understand the building as the organizing factor of our life's processes (*als Gestaltung von Lebensvorgängen*)? And who is to integrate the scientific, social, technical, and economic factors, inherent in the new architecture of the industrial age? Gropius' answer here is unambiguous. This integrative role is the essential task of the architect, and therefore of necessity he must be involved in the process of industrialized building of houses. Can the architect stand aside, asked Gropius, when even at that moment the combined forces of the German Steel Federation were united in an effort to "fill the land with steel houses"?[58] The theoretical framework constructed by Gropius for the enhancement of environmental and housing quality through prefabrication demanded the architect's engagement in the process, as a vital instrument not only of design but of total synthesis. If the architect did not step in, then, Gropius feared, industry would act alone. In this fear Gropius of course was fully justified. The pragmatic forces of economics were to prove a more powerful motivation for an attempt to industrialize the housing process than the logic and justice inherent in Gropius' elegantly formulated theories. It is to that development we must shortly turn. Before doing so, we should perhaps, in true Gropian manner, attempt a synthesis of Gropius' various theoretical statements as a summary of his principles of industrialized housing.

The architect, who is responsible to society, must contribute to the solution of society's most acute problem, the mass-housing problem. Mass housing demands mass production, and for this purpose the architect must learn to work with industry. Industrialization of the housing process inevitably means standardization; we should not resist this, for standards are the norms of a civilized community and give it unity of expression. However, we must not forget that individual needs and desires vary, and within the limits of social consensus man must be given choice. Man and his world are not static but in a state of dynamic flux, and the dwelling produced by industry must be adaptable and responsive to demands for change and growth. Industrialized housing must therefore be designed for maximum utility, standardization, and interchangeability of the parts and maximum variability of the whole, the house as final product. This industrialized building system moreover is not an end in itself but an integrated part of a larger whole, one level in a hierarchical environmental–social-economic system.

Gropius' theory of prefabrication was wholly consistent with his philosophy of life. If there is a theme that runs through all Gropius' work, it is the theme of wholeness, the theme of unity, the theme of synthesis.[59] His concern in every undertaking was to demonstrate the essential oneness of life, its organic unity whereby all parts constitute a whole, or series of wholes combine in new unions to form wholes of a higher order of structure, value, and purpose. It was from this point of view that he sought to resolve the dualisms of art and life, art and technology, in new syntheses. The philosophy underlying Gropius' theory of prefabrication is essentially of this unifying, synthesizing approach. On the surface this was not immediately apparent. In principle, Gropius had to choose between two main streams of development, in seeking to adapt industrial techniques to the processes of building: the industrial production of the complete building in a variety of predetermined variations, a process analogous to the mass production of the motor car, or the manufacture in factories of the components of buildings as a standardized set of elements with rules of combination allowing for variability of the final house design.

The application of a production-belt system to the fabrication of the total dwelling had obvious advantages not only in reducing the cost structure but in achieving a uniformity of design and quality, with a high degree of control over the whole process. The process is a determinate process of controlled synthesis leading to apparent maximization of unity. This was the direction Buckminster Fuller chose for his Dymaxion house; it was the direction one would have anticipated for Gropius, whose modus operandi was synthesis and whose aim was unity. Yet the import of his theory of prefabrication was to deny the validity of this approach. "Total" prefabrication was discarded as being only superficially consonant with his philosophy of unity, for it was antagonistic to his fundamental aim, the ultimate unity to which he strove, the integration of art into the life of man. The prefabrication of the complete dwelling was, in his view, based on a mechanistic world view. It was concerned with providing a given quantity of units in response to a need stated in numerical terms; it produced architecture for the statistical mean and not for the rich diversity which is human life. Unity produced by such means is an imposed unity which excludes free choice and personal preference.

Gropius turns therefore to the alternative avenue of approach and seeks to reap the advantages of prefabrication through the manufacture of the standardized part rather than the whole, leaving the synthesis to be defined by the needs of the user and the creative skills of the architect. Gropius' attitude to prefabrication is thus an attempt to resolve the conflict of uniformity and variation, of standardization and diversity, in a new unity: and it is typical of Gropius that he sees the resolution of this paradox not in a choice of alternatives but in a reconciliation of opposites in a new synthesis. As an instrument of this synthesis, as a means of achieving unity of the whole, standardization of the part becomes a consideration of prime importance. The synthesizing process depends on the nature and quality of the component parts. The demands of the technology of production, the demands of economics, the demands of the quality of design, and the demands of order and aesthetic unity all converge in the concept of the standardized component. Gropius willingly accepts the concept of standardization as culturally beneficient. "As a maturer and more final model than any of the individual prototypes merged in it, an accepted standard is always a formal common denominator of a whole period."[60] These standards, however, were not fixed and immutable but were in Gropius' words to be "constantly checked and renewed." The quality Gropius was seeking, through standardization, was to be "dynamic, not static, to serve as an inexhaustible stimulus to man."[61]

The practical experiments of Gropius were much more limited than the grand sweep of his theoretical vision. At Toerten-Dessau his unity is an imposed unity, achieved through the repetition of standardized, identical dwelling units. At the Weissenhofsiedlung we have the dwelling conceived as a static, finite object. These experiments, however, must be considered as means, not ends. They were conducted to examine the technical problems of industrialized building on the building site at Dessau, in the factory at Stuttgart. They were laboratory experiments under controlled conditions, examining limited hypotheses. At all times, however, through the reiteration of his broad theoretical principles, Gropius kept the wider picture and the bigger dream in mind. At no time did he regard the technical solution as an end in itself. The same, however, could not be said of the pragmatic forces of industry.

3

The Pragmatic Tradition

Industry as an Initiating Force: The Steel House

Responding to the assessed needs of the housing market in the mid-twenties, German industry—and not only the building construction industry—began, somewhat tentatively, to turn its attention to the field of industrialized building. The use of sophisticated steel structures, following American models, had not only been seen for some time in industrial and commercial buildings but by 1926 had also been introduced into residential buildings by the Luckhardt brothers and Anker in a private house, and by Mebes and Emmerich in the same year in a large apartment development at Berlin-Britz.[1] Now, consideration began to be given to the mass production of houses, using unorthodox materials and methods. "The shortage of dwellings in Germany is undoubtedly hastening the development of steel house construction there," advised the U.S. Department of Commerce, basing itself on a report, in 1928, from its consul in Cologne. This shortage, estimated at 800,000 dwellings, was "not without its effect in promoting the popularity of quickly erected, easily financed, durable, and hygienic quarters such as are just now becoming available to the public through the medium of steel houses."[2]

In addition to this perceived demand there was on the part of the steel industry and its associated branches an anxiety to exploit more fully a productive capacity that had begun to exceed marketing possibilities. "A big slump in the steel trade in 1927 left Vereinigte Stahlwerke, the German steel trust, with considerable excess capacity and the desire to seek new outlets,"[3] noted Burnham Kelly, adding that even major industrialists like Hugo Stinnes (a founding member of the trust) and Hugo Junker became interested in the potential of steel building methods.[4] Ambition and necessity combined to give a special impetus to the development of a new industry which, despite its innumerable permutations, was collectively called *Stahlhausbau*.

Steel houses produced by this industry were classified[5] in four classes: *Stahltafelbau*, a framed and sheathed construction, where the frame may not be steel but covered with steel plates; *Stahllamellenbau*, a structure of self-supporting steel panels that act as both structure and cladding; *Stahlskelettsbau*, a steel skeleton frame with an infill walling system that utilizes a variety of possible materials, including conventional masonry construction; and *Stahlrahmenbau*, a steel supporting structure preassembled into framed panels to simplify erection procedures. It is evident that not all of these methods are complete prefabrication systems; industrialization in some apply to the framing, in others to the cladding, and most involved a greater or lesser degree of site work, apart from assembly and erection procedures. But they were all far in advance of conventional housebuilding methods, and considerable savings in both cost and time of construction were hopefully anticipated. These inherent advantages, it was argued, would help to counter the resistance to the new products natural in a conservative society. As the American consul in Cologne reported: "The manufacturers of steel houses have launched considerable propaganda in their support in the Cologne district, hoping by these means to overcome the existing prejudice against their use and at the same time to increase their sale. This propaganda is most widespread, appearing in the local press in the form of favorable comment, through advertising, and also by means of free, illustrated lectures and discussions.'' These methods, we may interpolate here, were precisely those advocated by Gropius in the 1910 memorandum. The report continued: ''To date, the most effective arguments presented are those based on the simplicity of this method of construction, its practicability, low price, hygienic character, durability, and the rapidity with which these steel houses can be erected.'' He then added wryly: ''The prejudice against the steel house is traditional in the Cologne area and is based principally upon aesthetic grounds. The houses which are being erected are, admittedly, useful but cannot be called beautiful, which fact the producers seem to have taken into account in that they base their sales arguments rather on the practicability than the beauty of these structures.''[6]

In this new development precedents were sought and encouragement derived from experiments in other countries. The industry of Britain and America was still much admired and respected in Germany, and their achievements were looked on as a model for emulation. Therefore the extensive American experience with steel, in the structural skeleton generally and to a limited degree in the development of steel houses,[7] was regarded with interest. However, the more important influence was undoubtedly the British postwar experiment in factory-produced steel houses, especially of the Atholl, Braithwaite (Telford), and Weir types. These British developments, whose achievements were widely publicized but whose problems were not as widely known, were regarded as a great success in Germany. They looked with admiration at the range of types, at the eminence of their sponsors, and the thousands of houses reported to have been constructed. Most of all they looked to the Weir house, and its features were comprehensively described in German-language professional and construction-industry literature.[8]

The first German steel house of this postwar period, manufactured by the Wöhr Bros. Ironworks at Unterkochen, was generally acknowledged to have been derived from the Weir house.[9] It was produced in April 1926 as a five-roomed house of *Stahltafelbau* construction whose basis was a wood frame covered with 3-mm steel plates well insulated with *Torfoleum*. Later models used an alternative structure of steel framing. This more homogeneous system was also used by Braune and Roth of Leipzig, whose "Sonne" model steel house came shortly after the Wöhr house. Completed on 26 August 1926 it had taken a mere three weeks to erect, and eventually several one- and two-family house types were produced, all in a system similar to the Wöhr house, or its British prototype. Braune and Roth came to steel house manufacture from a background in the manufacture of steel strong rooms and safes; and their patented system of house construction was used by another Leipzig firm of strong room manufacturers, Carl Kästner and Co., who also began, at that time, to produce steel houses. It was Kästner who put together the Muche-Paulick steel house at Dessau, which we discussed in the preceding chapter. These pioneer efforts in the new steel house business originated not with architects nor even with traditional building contractors but with firms with the special equipment—the presses and mills—for processing steel and the knowledge and experience to adapt it to new purposes. Another such firm was the Hamburg machine tool manufacturer and shipbuilder, the Deutsche Schiffs-und Maschinebau-Aktiengesellschaft, and with its entry into the field of prefabrication, the wheel of history turned full circle, for nearly a century before some of the first iron prefabs had been made by the shipbuilders Fairbairn and Grantham.[10]

By far the most substantial industrial concern to engage itself in prefabrication was the newly established giant steel trust, the Vereinigte Stahlwerke, which set up a subsidiary company, the Stahlhaus-Baugesellschaft m.b.H., of Duisburg, "for the sole purpose of manufacturing steel houses."[11] These houses were of the *Stahllamellenbau* type, similar in some respects to the Telford house of Braithwaite; that is, they were constructed of load-bearing, self-supporting steel panels. An experimental house of this type was erected at Kaiserburg, in Duisburg, in 1927, and was followed shortly after by groups of houses in Duisburg and Munich. The system was developed for Stahlhaus-Baugesellschaft by the architect Heinrich Blecken, who was Vereinigte Stahlwerke's director of construction, and had been described in all its technical details by him as early as 1926 in an article in the *Deutsche Bauzeitung*, after the system had been patented.[12] When, some three years later, Blecken once again wrote of the steel house, in a comprehensive survey of *Skelettsbau* and *Lamellenbau*, he could point to many examples erected, including a complete steel house *Siedlung* in the Ruhr.[13]

One of the most important of the German steel house systems originated in neighboring Austria. This was the Böhler system, originally developed by the Viennese architect, Alfred Schmid.[14] He first reported on the new system in the journal of the association of architects and engineers in Austria in 1927,[15] but as he pointed out, the experiments he was describing had by then been proceeding for one and a half years, which would place them among the first pioneers. He acknowledges the contribu-

3.1
Wöhr Bros., steel house, probably at Unterkochen, 1926

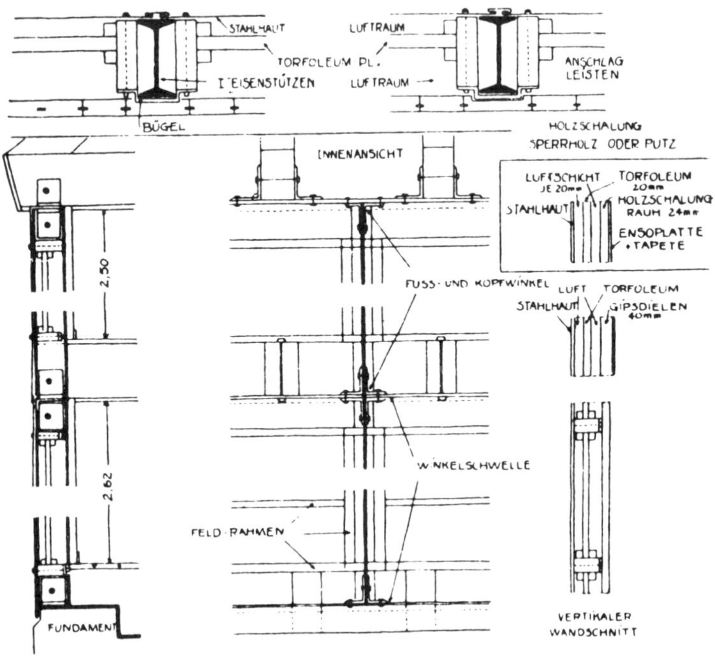

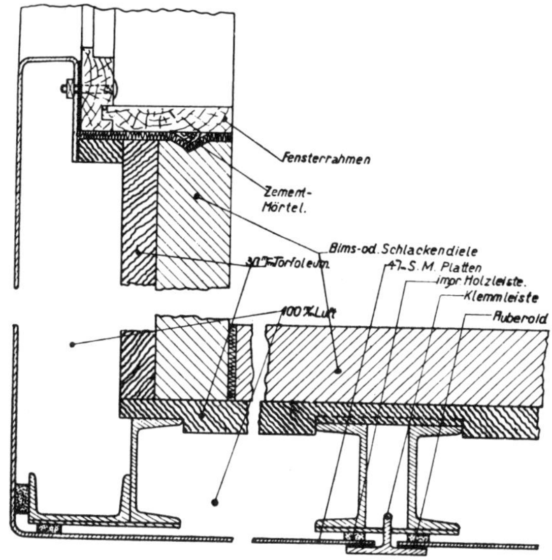

3.2
Braune and Roth, steel house, construction details, c. 1926

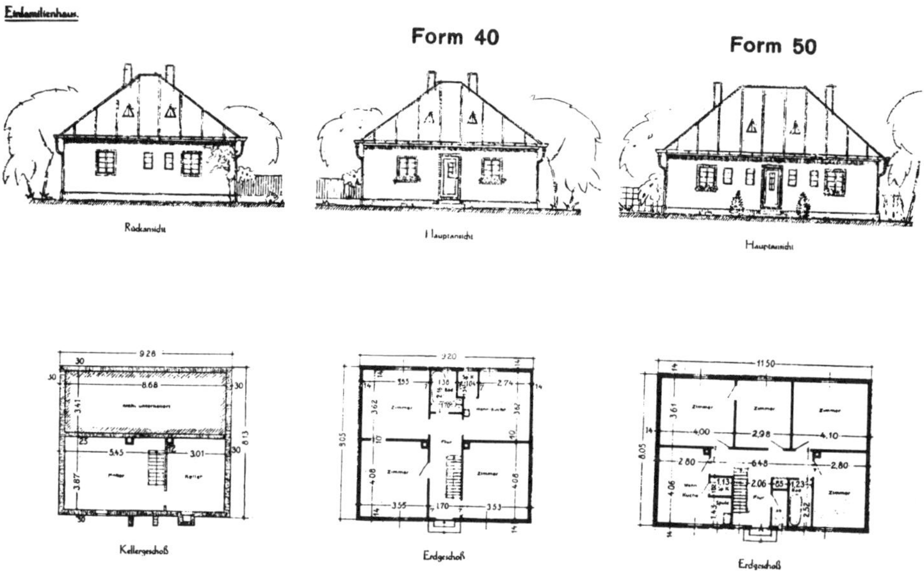

3.3
Heinrich Blecken, steel houses, Stahlhaus G.m.b.H., Duisburg, c. 1927

tion made by others to the development of industrialized building: the American experience in steel building, the Bauhaus experiments, and of course the Weir house, a major article on which had appeared in the previous issue of the same journal.[16] He views the industrialization of the building process with enthusiasm and with optimism. Its product is the "house of the future," and its day is not far off. The reasons he cites are familiar because they have become the undisputed assumptions of the new industry. The factory-made house must succeed, because of its economical cost, its speed and ease of erection, and its independence of the constraints of both the traditional labor force and the vagaries of the weather. The article sums up the achievements to date: an experimental house in Kapfenberg, and two houses at the exhibition Wien und die Wiener, one designed by the architect Walter Raschka and the other possibly by Schmid.

The manufacturers of these houses, Böhler Bros., are only briefly mentioned in the article, but they are the technical base and commercial power that motivates the entire operation. In 1926 the Gebr. Böhler & Co. Aktiengesellschaft of Vienna established their Abteilung Stahlhausbau to produce houses based on the Schmid system.[17] Several house types were developed, including one-family and two-family models. A branch of the business was established in Berlin to further sales, and was said in 1928 to be "active in this direction."[18] The Böhler houses utilized a steel skeleton frame of light channel sections at 1-m spacing, to which were bolted steel plates which formed the *interior* surface of the wall. The exterior was clad with Heraklith or other insulation boarding, finished with roughcast plaster. In later versions Böhler abandoned the inner steel plate, which had been its unique feature, and experimented with a variety of other materials, including Heraklith and Eternit asbestos sheeting as the inner cladding. Still other versions, while retaining the small-scale, closely spaced steel posts and beams (the enduring and essential elements of the system), replace the cladding with more substantial materials, such as pumice building blocks and slabs, cement rendered.[19] Although the structure of the Böhler house is prefabricated, in many of its forms its external finish is stucco, which makes it indistinguishable to the eye from conventional houses. The Böhler company made constant efforts to bring the advantages of its system to the attention of architects and the lay public. At the Bauwelt Musterschau of 1931 a house, designed by architects Kuhnert and Pfeiffer, using the Böhler system, was exhibited as part of the outcome of a low-cost housing competition.[20] Later the same year, the Böhler-Stahlskelett-Bauweise (as demonstrated in a house by Haring) was displayed to the public.[21] In the Houses for Fixed Prices exhibit organized by *Bauwelt* magazine the following year[22] (1932) there were five houses using the Böhler system, four designed by architects, the fifth a product of Böhler's own design department. The appearances of these houses are varied, and it is apparent that the use of the system imposes few constraints on the architect. This is also apparent in the diverse appearance of the Böhler houses designed by the architects of the Growing House, in the 1932 exhibition at Berlin, which we shall discuss later.

Gropius himself had for a long time taken a keen interest in the Böhler system. He had familiarized himself with their printed material, inquired of their conditions of payment and delivery, and was particularly interested in their use of flat roofs.[23] His own experimental house at Stuttgart, we recall, was based on a light steel frame, and as we shall see, he was to return to the idea again, developing it even further. The Böhler system, though essentially a compromise system of prefabrication, was one of the better known in Germany, and it continued in production until the mid-thirties, when restrictions on the use of steel

3.4
Walter Raschka, cottage, Böhler system, Wien und der Wiener exhibition, Vienna, c. 1927

3.5
Böhler-Stahlbau, schematic diagram, lightweight system, c. 1932

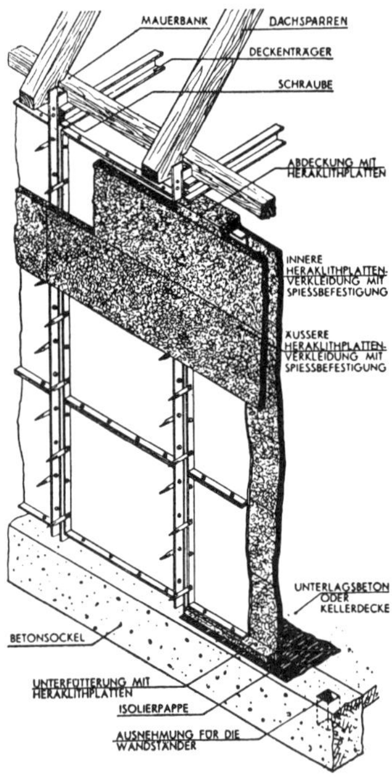

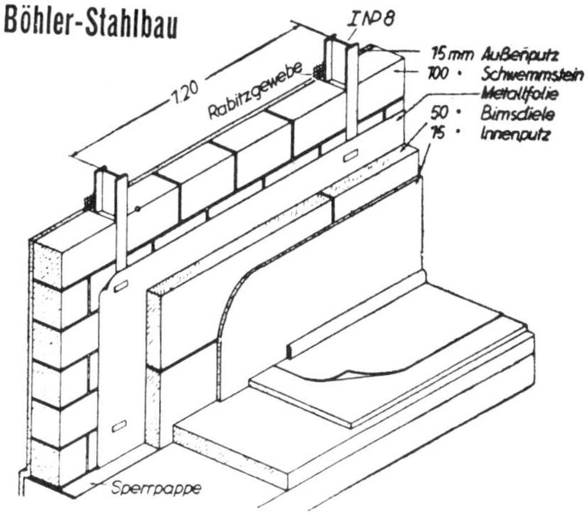

3.6
Böhler-Stahlbau, schematic diagram, block infill system, c. 1932

3.7
Johannes Niemeyer, "fixed-price" house, using Böhler system, Berlin 1932

forced it out of business.[24] As late as 1933 there were reports of a new *Siedlung* of 2,500 houses, to be built entirely of Böhler houses;[25] it was also, as we shall discuss later, actively to seek an export market. For a few years the steel house industry flourished in Germany where it was initiated very largely by the dynamics of the steel industry itself and its associated firms.

Konrad Wachsmann and Holzhausbau

Building in timber was a German tradition of venerable ancestry, which remained popular, even after the Industrial Revolution, in rural if not urban contexts. After the 1914–1918 war, as a result of a great shortage of all strategic building materials, timber was once again seriously considered, and its potential exploited, often in entirely new ways by architects and building contractors. The German construction magazine, *Deutsche Bauzeitung,* began to produce in 1920 a special supplement, *Der Holzbau,* to record and encourage this development. In its very first issue it featured an article by Adolf Sommerfeld on the role of the German Timber Building Federation, and an illustration of a large wooden hangar for naval aircraft, built by Sommerfeld's construction company with timber processed by his own sawmills.[26] Walter Gropius had at this time, 1920, just designed "a quite fantastic project for an administration building for the Sommerfeld company . . . completely of wood;" had, together with his partner Adolf Meyer, designed a two-story apartment house for Sommerfeld (built in Berlin-Lichterfelde in 1920–21); and was busy with the famed Sommerfeld blockhouse, which was a combined Bauhaus project.[27] Gropius was, in 1920, an ardent protagonist of wood as a building material for modern times: "We must re-experience wood, rediscover it, re-form it, according to our *own* spirit and without imitating old forms that no longer suit us.[28] Innovative (although romantically evocative) in form, Gropius' blockhouse for Sommerfeld was traditional, handmade, in construction. Sommerfeld himself, however, was developing a blockhouse building system on an industrialized scale, which was discussed and described in several issues of *Der Holzbau*.[29]

This activity by Sommerfeld was part of a general, but short-lived, revival of interest in timber building, whose climax, perhaps, were housing estates in Berlin of about 300 wooden dwellings, put up in 1920 by the Holzbau-Industriellen-Verbandes Berlin, a consortium of construction companies.[30] After this intense activity a decline set in, and after the spate of early projects in 1920 and 1921, by 1923 there is not *one* report of a new development in *Der Holzbau*.

In August 1926, in a lecture to the Association of German Master Carpenters, the engineer Leo Kuhberg addressed himself to the problem of timber building methods in the technological development of Germany.[31] He regarded timber as a resource whose potential was only recognized during the recent war, when shortages of steel and concrete necessitated a search for alternatives. After discussing the problems and possibilities of wooden construction, even in major engineering works, he turned to the issue of housing. This, for Kuhberg, was to be the main thrust of the economy and the building industry, to mobilize every resource to the last German mark for the building of houses. Here he foresaw a great role for wooden houses, because above all they responded to the call of the day for the establishment of norms and standard types. Whole walls, whole rooms could be premade complete to the last fitting according to these norms. Was it not ludicrous, he asked, that, with this great potential, housing authorities in the Rhineland and Westphalia were at that time busy negotiating with English firms for the importation of *Englischen Stahlpanzerhäuser*?

This *cri-de-coeur* reflected that diminishing of activity to which we have earlier referred, the decline the German wooden building industry was at that time suffering, both as the result of changed postwar circumstances and the competition of the technologically more sophisticated, more exciting, and more "modern" steel systems then being developed. In the field of one-family dwelling the traditional building industry focused on massive systems of masonry construction, and new initiatives, which were directed to the development of the steel house, were exceptional. But the wooden house building industry was temporarily eclipsed by these alternative methods. This situation was soon to change, and as we shall later see, the majority of systems at the Berlin Summer Show in 1932 were of wooden construction, with the major emphasis on prefabricated panel systems.[32] This renaissance stemmed in part from the revived interest of architects in wood construction, as they came to understand its potential in industrialized building. The extrapolation from traditional methods of detailing woodwork to the designing of total systems was not as great a step for a progressive architect as entry into the more esoteric field of steel construction. These initiatives by architects were underpinned, and given both breadth and stability, by the far more widespread and expanding activity of some of the long-established, substantial makers of wooden buildings. Let us briefly take a look at two of these firms.

Karl Schmidt was a cabinet maker who started a furniture workshop in Dresden in 1898.[33] This venture, at first called the Dresdener Werkstätten, later became the renowned Deutsche Werkstätten, whose main concern was to be the production of furniture by machinery. In 1905–06, at an exhibition in Dresden, the workshop showed to the public, for the first time, its range of machine-made furniture, much of it designed by Schmidt's brother-in-law, the architect Richard Riemerschmid. Riemerschmid was one of the original twelve artists who had been founding members of the Deutscher Werkbund, and the Deutsche Werkstätten had been one of the original representatives of industry. Karl Schmidt was a close personal friend of Friedrich Naumann, cofounder with Muthesius of the Werkbund. Another original member, the architect Adelbert Niemeyer, of Munich, together with a colleague, Karl Bertsch, now came to join the Deutsche Werkstätten group. Its aim, proclaimed the catalog at the Dresden exhibition, was "developing the style of furniture from the spirit of the machine."[34] By 1910 it had begun to concern itself with the issue of standardization and modular units which were developed as *Typenmöbel*.

The next step, from industrially produced furniture to industrially produced wooden housing, was logical and—considering the number of architects closely connected with the firm—inevitable. While the production of furniture continued, a new department for the production of "D. W. (or De We) Holzhäuser" was established. By the late 1920s this was a flourishing business, with headquarters at Hellerau, near Dresden, and with branches in Berlin and Munich.[35] They offered twenty-five standard houses designed by architects Karl Bertsch, Eugen Schwemmle, Richard Riemerschmid, Adelbert Niemeyer, and Hans Poelzig. These house types came in most cases in two alternative systems, finished externally either in *Jalousies* (weatherboards) or in *Schwartzen* (half-round timbers set in frames and treated with carbolineum). All wooden walls were thermally insulated and finished internally with smooth panels. Sizes of houses ranged from a modest 55 sq m (costing RM 8,960) to a luxurious 320 sq m (at RM 40,480). On the whole the architectural styles were varied but conservative, and there was no suggestion in the plan, the external form, or detailing that these were factory-produced houses. In fact they were much less like prefabricated houses than were a set of wooden houses, mainly flat roofed and modular, which were designed by Riemerschmid and published shortly thereafter.[36] The output of the firm was eclectic, and in addition they were prepared to adjust any design to suit individual desires. It was this quality they stressed—together with economy, speed of construction, good insulation, good fire-resistance qualities, and adaptability to landscape—in their campaign to sell the houses. They participated in competitions to bring their product to the public notice, and in *Bauwelt*'s modern small house contest an entry by Eugen Schwemmle, using the De-We system, was among the award winners.[37] Schwemmle was also the architect of their most impressive achievement up until the end of the thirties, the Grosssiedlung Leupnitz-Neuostra, near Dresden,[38] with both single and multifamily dwellings constructed in large numbers.

Probably the most important firm in Germany, and certainly the most venerable, to be engaged at that time in the industrial production of wooden buildings was Christoph and Unmack A. G., of Niesky, in Silesia. The history of the firm is complex and interesting.[39] Like Schmidt of the Deutsche Werkstätten, Christian Ferdinand Christoph began his career as the maker of furniture, in partnership with an architect, Christian Rudolf Unmack. This was in Copenhagen, in 1869. Some time later, in 1880, the firm of Christoph and Unmack began to participate in the development of a system of construction of wooden barracks, invented many years previously by a Danish military officer, J. G. C. Doecker. In its original form this had been more of a rigid tent than a building, comprising a system of wall and roof panels, each consisting of a wooden (or iron) frame covered with a waterproofed felt material, the panels being hinged together in pairs, for easy folding, stacking, and transportation.[40] When, in 1882, Christoph and Unmack acquired the patent rights to the Doecker-Bauten system, they developed it further into a more substantial structure and made provision for doors and windows, more permanent cladding and roofing materials, and an ingenious but simply operated metal coupling system using hooked catches, bolts, and wing nuts to secure and unite the component elements. These, it should be noted, were of standard modular size. The system was reversible, and the essence of the structure was not only its ease of erection but its demountability and transportability.[41] With the financial backing, technical ability, and initiative of this firm behind an idea basically sound in conception, the project flourished and soon became internationally known. A Gold Medal was won at the Berlin Hygiene Exhibition of 1883 and an award in an International Red Cross competition for Portable Hospitals in Antwerp in 1885.[42] These accolades stimulated trade in the field hospital and barracks buildings made according to the Doecker system. In 1885 an order for 50 barracks was received from the Prussian Ministry of War, and this proved the decisive factor in persuading the manufacturers to move from out of the way Denmark to the heart of Germany. They came to Niesky in Silesia, where half a century previously another Christoph, J. E. Christoph (a cousin to Christian Ferdinand) had established an iron machinery and construction business. It was by now a flourishing concern, run by an engineer son, and employing several hundred workmen. In 1887 the two businesses joined forces, and a new Christoph and Unmack emerged at Niesky, a conglomerate factory

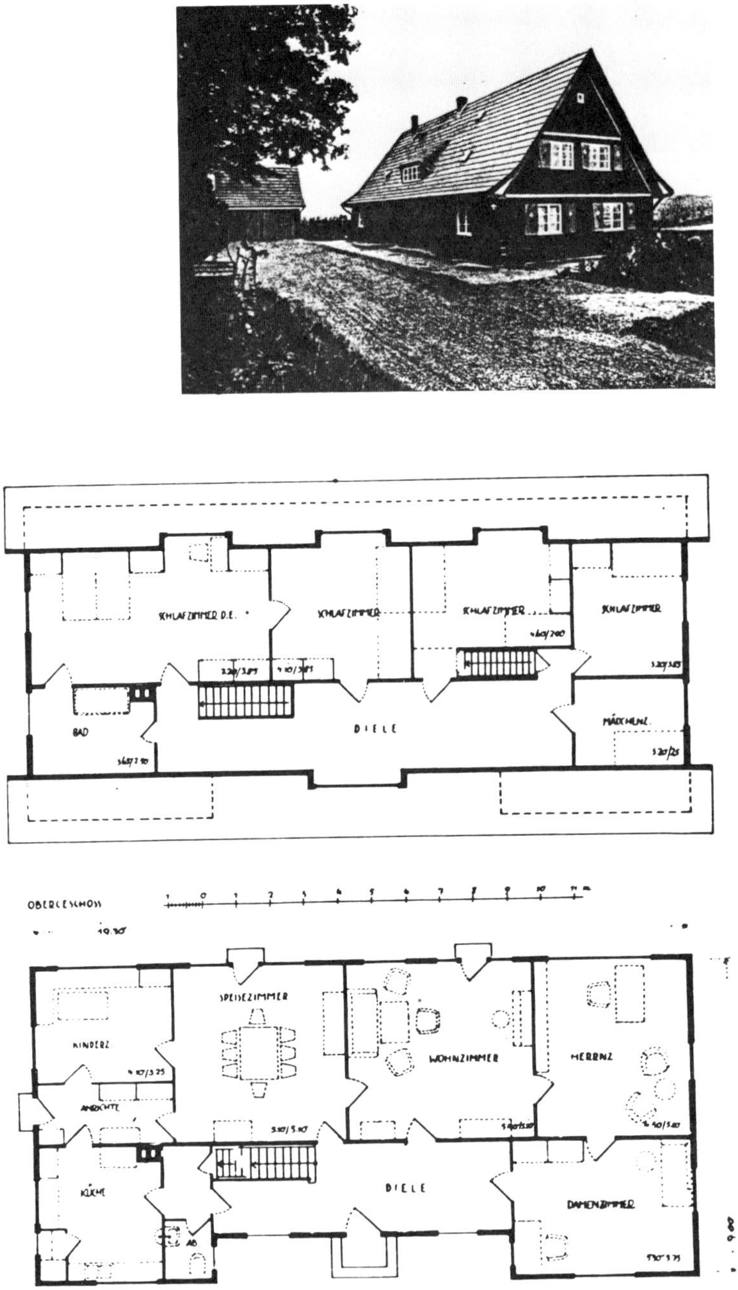

3.8
Hans Poelzig, house using De-We construction system, c. 1929

1 : 300

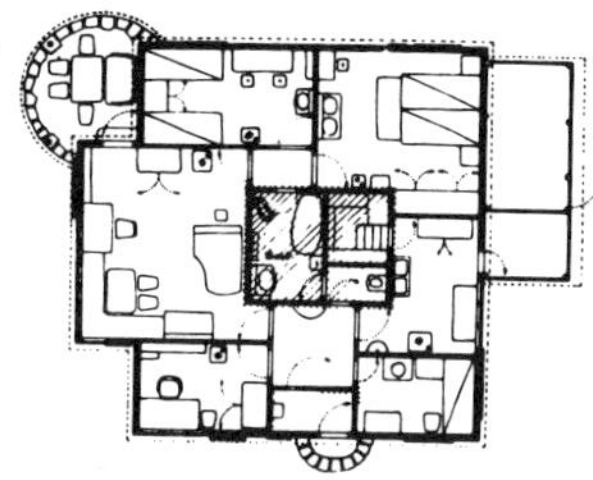

Bild 22 und 23 / Holzhaus mit Wohnraum, Arbeitszimmer, zwei Schlafzimmern, Schlafkammer (5 Betten), Küche, Bad, Abort, Keller und Schuppen

1 : 300

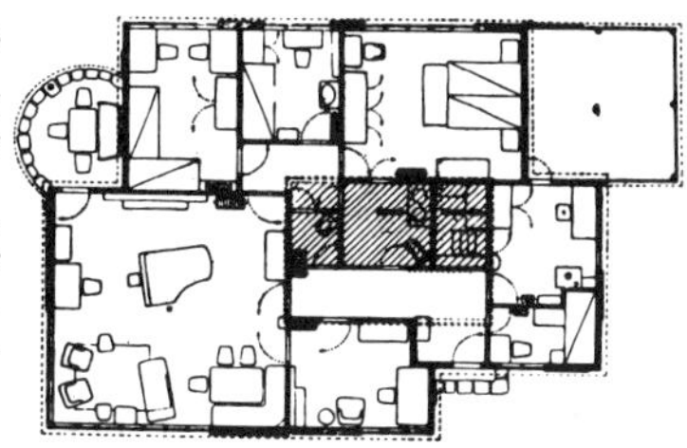

Bild 24 u. 25 Holzhaus mit Wohnraum, Arbeits-Zimmer, drei Schlaf-Zimmern, einer Schlafkammer (sechs Betten), Küche, Bad, Abort, Vorratsraum, Keller und Schuppen

3.9
Richard Riemerschmid, designs for wooden houses, 1932

3.10
Eugen Schwemmle, Leupritz-Neostra housing estate, precut timber houses

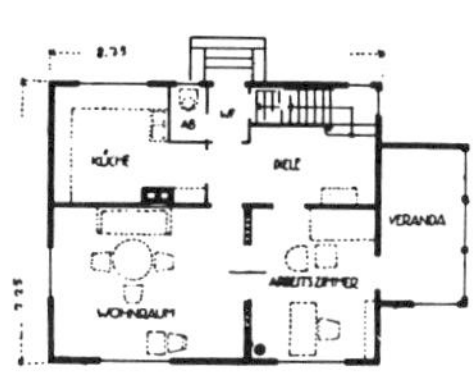

Grundriß vom Erdgeschoß

manufacturing machines, ironwork, and wooden houses. By the beginning of the twentieth century it was already a large and flourishing business, making a large diversity of wooden buildings. World War I, with its insatiable demand for barracks and field hospitals, gave production an additional impetus. At the war's end it continued this military production on a diminished scale and turned its attention to civilian needs, seeking to capture markets other than emergency ones. It sought to improve its quality and, in 1919, initiated scientific tests to enhance the thermal performance of its blockhouse and panel systems.[43] By 1920 it had begun to build wooden houses in significant numbers, in the new *Siedlungen* near Berlin, at Johannisthal and Berlin-Steglitz.[44] Illustrations of its specialized buildings: hospital pavilions, a pilot's home, portable gymnasiums, even a school installation exported to Leith, in Scotland, were published.[45] After this promising new development came the economic decline, and Christoph and Unmack went through the troubled years of 1923 and 1924, years of financial chaos and social disorder, with great difficulty. By 1925, with surprising suddenness, the economy revived, and with it the fortunes of this giant and diverse manufacturing company which in its departments for wood construction, machine construction and wagon building employed some 4,000 workers, with an administrative staff of 400.

In 1925 a new director, the engineer Friedrich Abel, took over the wooden building department, and a vital new period in the firm's development commenced. The architect Hans Poelzig was appointed to the Board of Directors in 1926, and although he did not introduce revolutionary changes, Christoph and Unmack was exposed for the first time to the new world of architecture then unfolding in Germany. Through Poelzig there was a direct line of contact with the group of architects who were then remaking Germany's cities and its architectural character.

The eminent Hans Scharoun, for instance, was commissioned to design a "middle-class" house in 1927. Poelzig's contribution to Christoph and Unmack lay not in this, nor in his own rather undistinguished designs, but in another direction altogether. Heinrich Wurm puts it this way: "Poelzigs eigentlicher Beitrag bestand freilich darin, dass er seinen Meisterschüler Konrad Wachsmann als Chefarchitekten nach Niesky vermittelte" (Poelzig's essential contribution of course was that he established his master student Konrad Wachsmann as chief architect at Niesky).[46]

When we recapitulated Walter Gropius' early experience in prefabrication, we could feel a certain inevitability, a consistency, in an evolutionary process from first theoretical formulation to final full commitment to the industrial system. With Konrad Wachsmann, very much younger and much more ebullient in temperament, we sense a more accidental process at work. In his relationship to prefabrication, as in so many other critical points in his career, chance takes a hand, and when in 1926 he is eventually brought by the vagaries of fate into direct contact with the potentialities of prefabrication, it is a dramatic but unexpected confrontation.

Konrad Wachsmann came only indirectly to architecture.[47] As a youth he was apprenticed as a cabinetmaker and carpenter, and later became a skilled journeyman in this craft. In 1922 his latent interest in architecture found its first formal expression in studies at the School of Fine Arts in Berlin. During 1923–24 he took a course under Heinrich Tessenow at the Academy of Arts in Dresden; and the following year was a "master student" under Poelzig, at the Academy of Arts in Berlin. This was to be a most significant encounter, for a warm master-disciple relationship was soon established. When Wachsmann returned to Berlin from an abortive visit to Paris in 1926—

3.11
Christoph and Unmack, patent for military tent (German patent 1884)

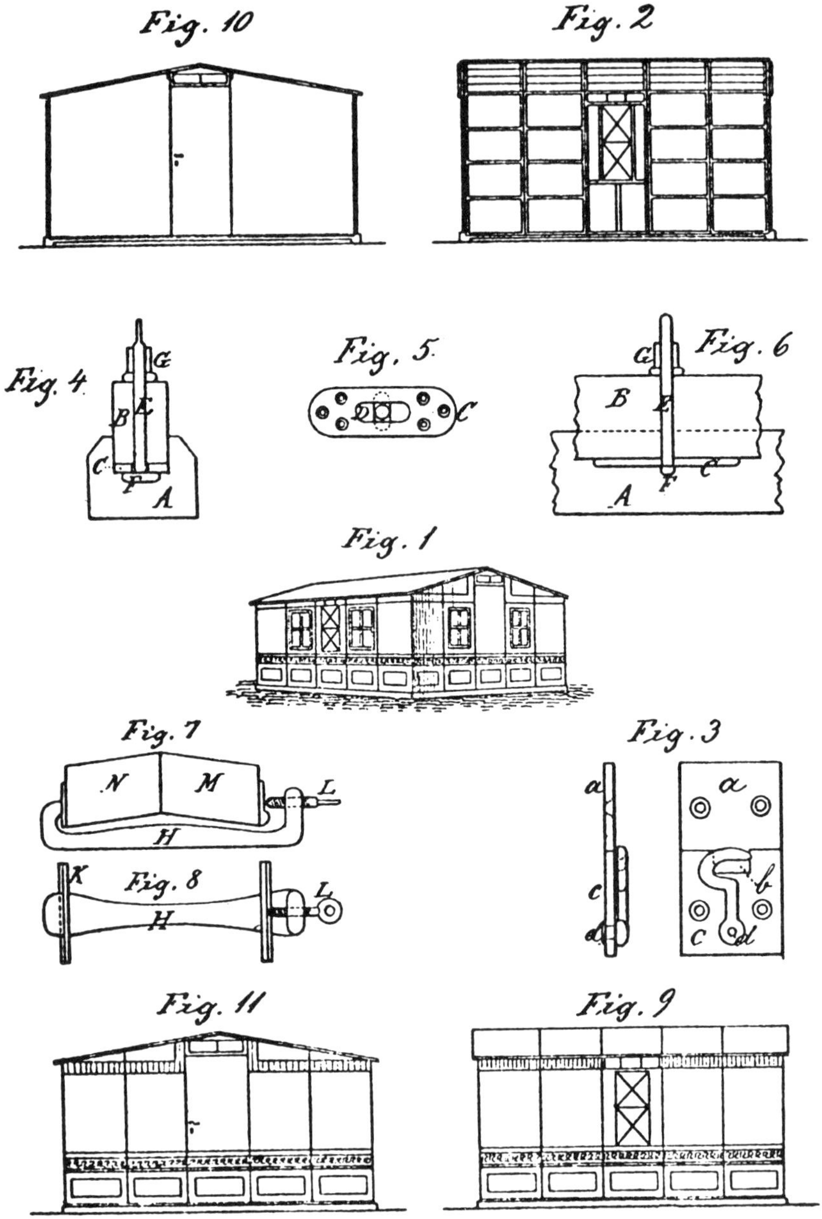

3.12
Christoph and Unmack, Ducker's (Döcker's) portable barrack and field hospital, 1886. *Top*: general views; *bottom*: details.

penniless, jobless, desperate—it was Poelzig who came to his rescue, with the offer of a job at the factory of Christoph and Unmack, where he was now a director.

Christoph and Unmack was at this time probably the largest and oldest manufacturer of prefabricated wooden buildings in Europe. It had branch offices in several major German cities and agencies in many more. It was also represented widely abroad, with agents throughout Europe as well as the Middle East and Latin America. According to the original Doecker panel system they manufactured single- and multifamily houses, and entire housing schemes; workshops and office buildings; exhibition and sales pavilions; children's, recreation, and convalescent homes; and standardized classrooms and school buildings.[48] In their blockhouse department they constructed villas, country houses, and hunting lodges of traditional Scandinavian appearance but utilizing ingenious prefabrication methods. In addition to these "transportable" buildings they undertook major engineering works in timber, ranging from industrial buildings to long-span bridges. Finally, they were engaged in the series production of joinery work, such as standard doors, windows, and staircases.

Although the blockhouse section produced a large variety of building types—the catalog of that department alone running to 67 pages[49]—it was the panel system that was technically and architecturally the most interesting.[50] It comprised a simple modular set of self-supporting timber-framed elements, ranging in width from 1 to 1.33 m, among which were standard wall, door, and window panels of various types. There was a variety of possible internal and external cladding specifications, as well as degrees of thermal insulation, depending on the class and purpose of building ordered. These panels, in some instances, appear to be complex in construction, over-designed, and extravagant in the use of material. This, however, was typical of much Scandinavian design and seemed also to be representative of German design in timber of that period.[51] The panels were joined together by metal catches (*Hakenverschlüsse*) let into the framework, four on each side, and the joints were normally concealed by vertical cover strips. This panel system, with its rigid modularity, generated a simple architecture of repetitive character capable of extension vertically and horizontally on an incremental basis. It was a modest, orderly, systematic architecture at once expressive of its mechanical origins and keeping with the spirit of the Modern Movement.

Christoph and Unmack used the services of many architects, in the design of their buildings. To Poelzig and Scharoun, whom we have already mentioned, we could add such names as Professor Albinmüller, Franz Zell, Joh [ann?] Mundt, Werner Schenck, Hans Herkommer, Hans Zimmerman, and many others. Most of them designed houses using the blockhouse system. Other architects[52] used the Christoph and Unmack framing system, which gave greater freedom, as it did not express the modularity of the components on the exterior but only in the interior wall linings. In relation to the panel system, however, with its greater discipline, the basic design work was generally done within the organization itself, in Christoph and Unmack's own design department. Wachsmann entered the design department of Christoph and Unmack in July 1926.[53] By April of the following year he had risen to the rank of chief designer and had achieved a position of much authority, being concerned not only with design but also with problems of research, management, and production. He was responsible for the design of several important projects: a house for the director, a convalescent hospital, and a hospital and school pavilions. Although Wachsmann made no changes to the construction sys-

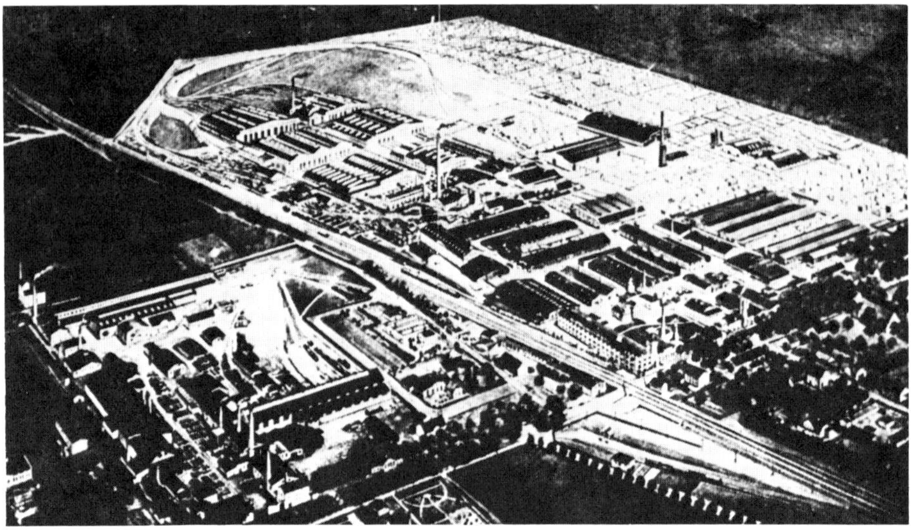

3.13
Christoph and Unmack, factory at Niesky, Silesia, c. 1926

3.14
Fritz Marcus, wooden house in Rupenhorn, Berlin, using Christoph and Unmack system, c. 1923

tem itself, he undertook the reorganization of the firm's catalog on a modular basis, providing a squared grid upon which customers could plot out their own plans. This aid to user participation was, incidentally, later to be used by Hans Scharoun in his project for the Growing House exhibition in Berlin, in 1931–32. "I developed new types of catalogs," wrote Wachsmann, "which I believe for the first time in Europe did not offer finished buildings but instead all components to build with. Modular grids had been printed in those catalogs in which clients could draw their own approximate floor plans. Those were then transformed into professional drawings by my office, using only numbered predetermined parts to build the whole."[54] In this way the emphasis changed from the production of a range of building types to a set of standard building components which could be freely combined according to the purchaser's needs. The importance of this concept must be stressed because Wachsmann had achieved in practice and at one stroke the goal toward which Gropius had been striving since 1910, that is, the standardization of the part and the infinite variety of the whole. Gropius, between the 1910 memorandum and the Bauhaus experiments, defined the theoretical postulates; Wachsmann, possibly unaware of the theory, achieved the practical consummation at Niesky. Notwithstanding this emphasis on the part, every building to leave the factory (components packed in crates which could also serve as floor panels) was fully equipped in every way: prefabricated foundation piers were supplied, electrical and mechanical equipment fully installed, and all necessary furniture included, even—so Wachsmann maintained—the chalk for the school blackboard and the surgical instruments for the hospital!

In 1929 Wachsmann, in a dazzling display of that mixture of initiative and chutzpah so typical of the man, managed to secure the commission to design the residence of Professor Albert Einstein in Potsdam. Using the full power of Christoph and Unmack's design department—draftsmen, engineers, accountants—he produced, literally overnight, an overwhelming presentation of the scheme: sketch plans, fully detailed drawings, mechanical and electrical installation diagrams, specifications, cost analysis. After some initial difficulties in finding a site, the project went ahead. The house was manufactured in Christoph and Unmack's workshops, using not the standard panel system but an alternative framing and cladding system sometimes employed. The finished house was first erected in one of the factory halls to test the fit of the components, as was Christoph and Unmack's practice,[55] and was then dismantled and transported to the Potsdam site for final assembly. The Einstein house is not a central event per se in the history of prefabrication, but it has a place in our narrative, and not only because of its piquancy. It was the occasion for Wachsmann to leave the protective environment of Christoph and Unmack and to enter the exposed field of private practice. But although for the time being he had given up his active role in prefabrication, most of his own designs being based on the more individual timber-framing method than on the universal panel system,[56] his view of architecture and technology had been irrevocably changed by his experiences at Christoph and Unmack. As he later put it: "In the large factory halls I saw for the first time, like a miracle, production machines producing . . . prefabricated panel systems for housing, hospitals and schools, manufactured there . . . and then shipped all over the world. In a split second I understood that mass production was more than a technological event. In fact, I suddenly sensed that industrialization was the answer to building, and terribly important." Wachsmann, who considered himself then, wryly, to be "only a carpenter," realized not only the limitations of the individual but the tremendous potential of working through the reproductive capacity of the machine. "This," he proclaimed, "was my revelation."

3.15
Christoph and Unmack, standard classrooom, c. 1930

3.16
Konrad Wachsmann, house for Albert Einstein, Potsdam, using Christoph and Unmack system, 1929

The lessons learned at Niesky, were spelled out by Wachsmann in the foreword to his book, *Holzhausbau*, written in 1930:

Today the wooden house is produced by machines in the factory, and not by manual labor. The old, highly developed, manual skills are being replaced by machine technology. In this way new possibilities are discovered, new applications and forms. Wood as a construction element formed through the traditional work of the carpenter no longer meets the requirements of fabrication and structural design. However, as a consequence of machine processing, it possesses the same technical and economic significance as any other construction material.[58]

In this book Wachsmann sums up his experience in wood construction and systematizes it in three modes: framing systems, panel systems, and blockhouse systems.[59] Each method is discussed, copiously illustrated, and expressed graphically in a sectional perspective of almost ideogrammatic lucidity. The examples illustrated in the book are derived from three main sources: Christoph and Unmack, the Deutsche Werkstätten, and the Carl Tuchscherer Co., all three pioneers in the industrial production of timber houses. Many of the illustrations are of buildings, in all modes of construction, designed by Wachsmann himself. This lack of modesty was characteristic of the man; also characteristic, however, was his generosity, for the book was dedicated to Friedrich Abel, the director of Christoph and Unmack, to whom he owed so much.

The Role of the Architect

From this brief survey of the German prefabrication industry of the 1920s it is clear that the architect played a varied role. In some instances the architect himself was responsible for the design of the prefabrication system, in its conceptual and technical aspects. If such a design resulted in a concrete consummation, it was generally in the form of a unique, one-off example, hand built by a construction firm rather than the product of an industrial process. It was an experiment in new construction methods, a demonstration of industrial potential, at best a possible prototype for industrial production. This, essentially, was the meaning and value of Gropius' demonstration houses at the Weissenhofsiedlung; and later at the Berlin Building Exhibition of 1932 it was to be the principal function of the work of many architects, including Scharoun, Ludolf von Veltheim and Klaus Müller-Rehm, Hans Köhler and Jürgen Schweitzer, and Martin Wagner.[60] Rarely indeed was the next step attempted, to try and put the design into actual production. Otto Bartning created his own company, Werfthaus System Bartning G.m.b.H., to produce his ingenious panel system; and Gascard and Canthal produced their own "G. & C.-Anbauhaus."[61] These were exceptions and did not prove to be commercially viable. Occasionally, an architect-designed system was adopted by an industrial firm, as was the case when Böhler produced the system designed by Schmid, or when Hans Spiegel's designs were produced by Stahlbau Düsseldorf G.m.b.H., in their Stahlrahmen houses.[62]

Much more frequently the systems were designed within the industrial organizations themselves by anonymous designers who might have been architects but were more likely to be engineers or even technicians familiar with the materials, machinery, and processes of production. In practically all the larger and more successful undertakings, whether in steel or in wood, the architect's role was principally that of an external consultant, designing the *houses* but not the *system* itself. We have already referred to many examples in relation to Böhler, the

Deutsche Werkstätten, and Christoph and Unmack. Some of these houses remained one-off examples for specific clients, others were demonstration or competition houses,[63] while many others became prototypes for mass production, being listed as standard types in the catalogs of the firms. In working out these designs, in terms of a specific existing system, it was possible that the system itself underwent modification. This was probably the case when Muche and Paulick adapted the Kästner system for their steel house, or when Josef Hoffmann designed his elegant steel house using the Vogel and Noot system.[64] Hans Scharoun modified the Christoph and Unmack panel system, using some standard wall panels horizontally, above and below strip windows.

Rarely do we have an instance of an architect operating, as Konrad Wachsmann did, within the industry itself. Perhaps the most notable case is that of Heinrich Blecken, the architect who acted as Vereinigte Stahlwerk's director of construction and who had, as we have already seen, been instrumental in the technical design of the prefabricated system from its very inception. Wachsmann could not of course play such a seminal role. When he came to Christoph and Unmack he was entering into a well-established firm, producing buildings according to long-held, patented, designs. His task, within Christoph and Unmack, was not to initiate a constructive system but rather to refine and develop it, and particularly to correlate techniques of production with concepts of design. In this his position was unique among all the noted architects interested in prefabrication. He eventually came to formulate a theory of industrialized building,[65] a theory whose sophistication is only faintly foreshadowed in his *Holzhausbau*.[65] This is no a priori theory, however, rather a crystallization of his years of practical experience, especially in the drafting rooms and workshops of Christoph and Unmack. If he moved toward a theoretical position, it was from a solid pragmatic base.

With Gropius, the reverse situation obtained. From the very outset he proposed a clearly articulated theory of prefabrication. Although this was enriched and refined over the years, he had not, as yet, been able to underpin theory with sound practical experience, other than desultorily in his isolated experiments. This situation of course was not of his own volition. In principle he had always advocated a simultaneity of theory and practice: this was the very essence of his educational approach at the Bauhaus, and it was his goal in practice. In this he had been constantly frustrated, ever since, in 1910, he had failed in his approach to AEG to achieve an industrial base from which to launch his venture in prefabrication. Ironically he remained cut off, by the exigencies of uncontrollable circumstances, from the real world of commerce and industry, despite a theory which, as we have seen, consistently called for the social responsibility of the architect, the need to wed theory to practice, and the ultimate synthesis of art and industry. In terms of this comprehensive philosophy Gropius sought a more durable and substantial alliance with industry. In 1928 he had talks with his old friend and former client, the building contractor Adolf Sommerfeld, who, as we have already noted, was something of a pioneer in the development of the *neuzeitliche Bauweise* in timber, about collaboration in the rationalized production of houses.[66] As an outcome of these negotiations he was sponsored by Sommerfeld to visit the United States and examine the nature of its building industry. He came back from this visit deeply impressed by the efficiency of American building, and with a great deal of information on systems of control and organization.[67] The practical outcome of this proposed business venture is not clear. There is some evidence that a

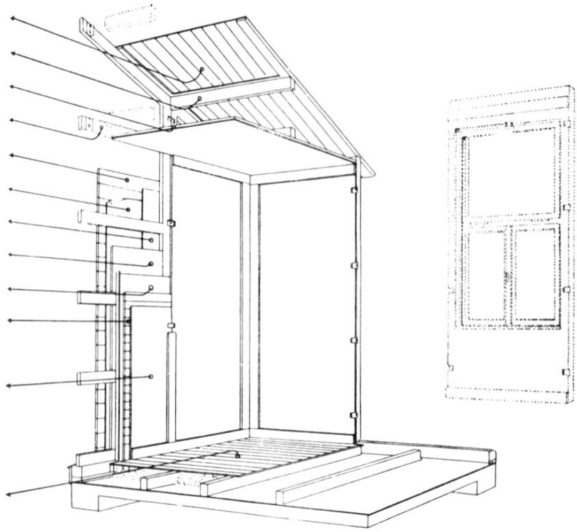

3.17
Konrad Wachsmann, diagram of Christoph and Unmack type panel system, c. 1930

3.18
Josef Hoffman, steel house, made by Vogel and Noot, Austria, c. 1929

3.19
Hans Scharoun, house using Christoph and Unmack panel system, 1927

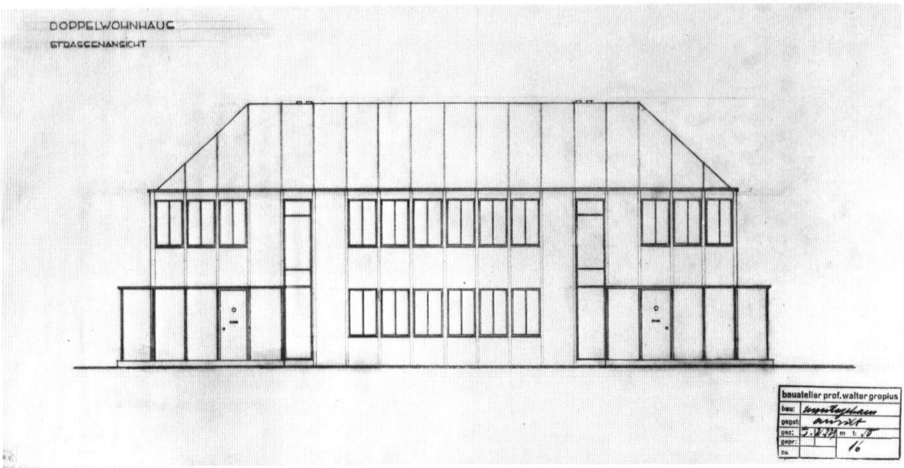

3.20
Walter Gropius, two-family dry-assembly (*Montagehäuser*) house, 1929. *Top*: street elevation; *bottom*: perspective.

few hundred houses were erected by Gropius and Sommerfeld, in Berlin and Posen,[68] but this may have been just prior to the American visit when Sommerfeld erected the Fischtalgrundsiedlung in Berlin, a project in which both Wagner and Gropius were intimately concerned.[69] Gropius' drawings contain a set of more than 30 sketch plans, details, and rendered perspectives, of one- and two-family *Montagehäuser*, prepared in March 1929.[70] These are of light steel frame construction, with aerocrete infill panels, deriving perhaps from the Weissenhof experiment, and bearing some resemblance to the Böhler system. The system was illustrated in both flat and pitched roof variants. We have no formal indication of the purpose of these proposals, but it is perhaps legitimate to speculate that they were for the proposed Sommerfeld venture, despite the fact that they were not to be executed in timber. Despite the considerable amount of effort invested by Gropius in the design and detailing of these houses, the venture apparently did not reach any practical consummation, for there is no evidence that any of these houses were ever actually constructed, even in prototype form—in fact they do not appear in the standard catalogs of work in the Gropius literature to any substantive degree. Nevertheless, this experience was not entirely wasted. Two years later it was influential in shaping the houses Gropius was to design, in his most serious and sustained encounter with industry, for the Hirsch Copper and Brass Works.

4

Gropius, Hirsch, and the Saga of the Copper House

The Hirsch Copper House

The Hirsch Copper and Brass Works (Hirsch Kupfer-und Messingwerke) were founded by Aron Siegmond Hirsch in 1906, as the extension and consolidation of an old-established family metal business based in Halberstadt, with associated enterprises in Werne, Ilsenberg, and Eberswalde. The firm developed to become a major power in the German copper industry and is described as having played "a leading role in German economic life."[1] (It was at Eberswalde, incidentally, that Hirsch built a large two-story new factory, designed by architect Paul Mebes, and later published by Walter Gropius in his *Internationale Architektur*.)[2] Hirsch not only dealt with copper ore but also with the manufacture of copper products, including such building products as copper tubing, sheeting, and roofing.[3]

During 1930 the Hirsch company began to experiment with the use of copper in building on a more comprehensive scale. They acquired the rights to a system of prefabrication of dwellings, invented by Friedrich Förster (originally Frigyes Förster, of Budapest) and later further developed by Förster in conjunction with Robert Krafft.[4] Förster, in his original patent application of 1924,[5] drew attention to the many previous attempts to design "knockdown buildings that can be readily assembled," which had failed because of high costs or through inadequate standards of construction and performance. He then went on to claim:

Recognizing the importance of such considerations I have constructed a knock-down type of building composed of structural elements that are made at the factory in the desired form and to the desired dimensions required by the purchaser. Each structural element constitutes a wall section of box-frame construction, and is adapted at its edges to be joined to other sections to provide a complete wall.

The wall sections may be made up of wooden skeleton frames covered on both sides with metal sheathing. The space between the metal sheathing is filled with insulating material such as wood wool, sawdust, exelsior, or the like. The edges of the sections may carry tie bolts or equivalent forms of fastening means for easily connecting the sections together.

The general construction of the sections is such that they can be readily made up at the factory and transported to the point of construction.

The advanced nature of these proposals must here be stressed, as well as their early date, in a European context. They were developed at the very latest in early 1924, and possibly in the previous year, and therefore predated all the German steel houses by at least two years. The Förster concept moreover had no structural prototype among the German houses, or even the earlier, much-admired British models. His self-supporting metal-faced panels eliminated the need for a structural frame, but without the excessive weight of the Telford panels and with a superior degree of thermal insulation. Moreover—and here lay the most significant contribution—the entire panel could be factory made, reducing the site work to a minimum.

In 1930 Förster, now working with Krafft, radically altered his concept of insulation, and a new patent application was made for a revised system that did away with the infill of insulation material and replaced it with a series of ''parallel partitions of a material which is impervious to air, preferably metal.'' This principle, incidentally, had already been exploited in practice in the panel system of the wooden houses produced by Holzbau A.G., a firm of house manufacturers operating under the direction of the architect Kunz, and had been published as early as 1922.[6] The performance could ''be further improved by lining the partition with a layer of a poor heat conductor, preferably of a fibrous material,'' in order to impede vertical air movement within the panels and to eliminate the possibility of heat bridges from exterior to interior. Dramatic improvements in thermal capacity were achieved, many times that of an equivalent brick wall. Hirsch was later to claim a thermal equivalence to a 220-mm-thick masonry wall for the panels they manufactured according to this system. In Förster and Krafft's patent application, a revised and much improved method of joining the panels was suggested in place of the original, rather primitive tie bolts. The wooden frames were beveled and bolted to a continuous U-section connector. By this device two-, three-, and possibly four-way connections were easily achieved without altering the standard panel, and at the same time heat bridges between adjacent panels were avoided. A universal jointing system was thus proposed. At the time of the revised patent application, in August 1930, Förster and Krafft gave their address as at Finow, Mark. This was the location of the Elberswalde factory of Hirsch Kupfer-und Messingwerke. Hirsch set up a sophisticated production process for the manufacture of houses on the Förster and Krafft principle, including assembly of the subcomponents to make building elements on a moving production belt, and a new division was established in the factory, the Copper House Department. A five-roomed model house war erected probably late in 1930, and it generated much interest not only in Germany but also in the United States, where it was studied with care by the Copper and Brass Research Association and published in the metal industry journals.[7]

In order to publicize this new venture, a catalog was printed, and a total of six examples of complete houses were exhibited, some at the Paris International Colonial Exhibition of 1931 (winning a "Grand Prix") and others at the German Building Exhibition in Berlin in May of the same year.[8] The "Haus in Allkupfer-Bauweise der Hirsch Kupfer-u-Messingwerke A.-G. Berlin" as exhibited was a large, two-storied structure made of wood-framed panels covered externally with copper sheets, internally with pressed steel plates, and insulated in the prescribed manner. The ceilings too were of pressed steel, suspended from wooden trusses, ready cut (with the parts numbered for ease of assembly), which carried the copper roof. The cost of the Berlin exhibition house was given as RM 10,900, and an erection time of only 24 hours was claimed.

Reaction to these exhibition houses was mixed. The editor of *Bauwelt* magazine in a private letter[9] confessed to some disappointment. He was worried about such technical problems as the heat conductivity of the copper panels; he anticipated long-term maintenance problems, particularly if the houses were erected in areas where the necessary skills and materials were lacking. He also pointed out that the price differential with conventional housing was not sufficient and—perhaps most important of all—the architectural character (more picturesque than functional) set back the cause of modern architecture by

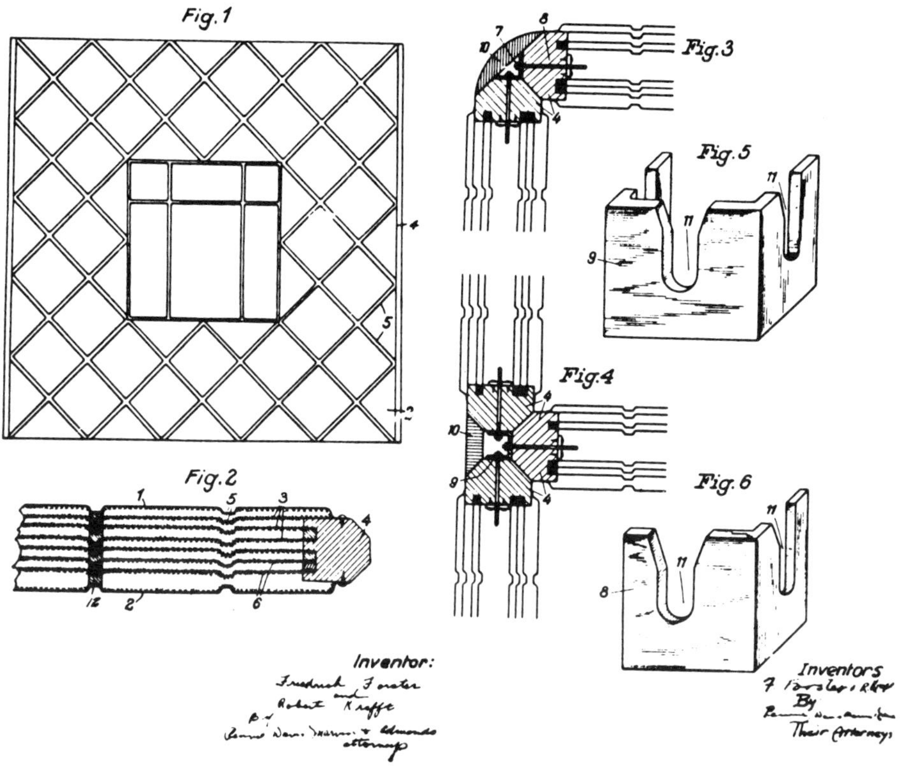

4.1
Förster and Krafft, heat-insulating wall (U.S. patent application, 1930)

4.2
Hirsch Kupfer-und Messingwerke, copper house, 1931.

4.3
Hirsch Kupfer-und Messingwerke, copper house, German building exhibition, Berlin, 1931

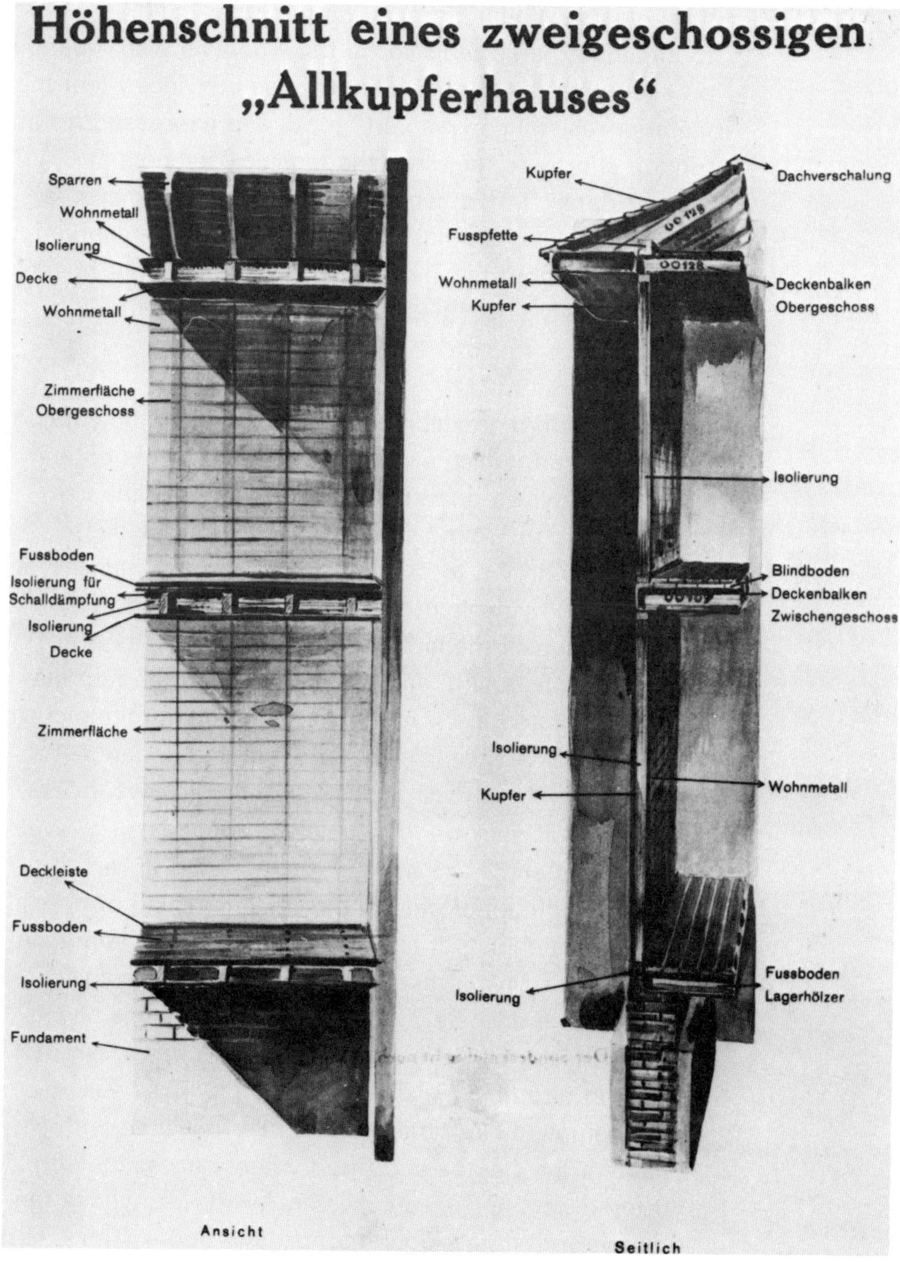

4.4
Hirsch Kupfer-und Messingwerke, construction details, 1931

thirty years. Walter Gropius, on the other hand, in reply to an inquiry, affirmed that on the whole he was most impressed with the copper houses, was convinced that their weather-resistance was very good, and expressed his intention shortly to look into the technical aspects in greater detail.[10] This general approval, as we shall see, did not preclude Gropius from having his own reservations, both about technical aspects of the copper houses and especially about their conservative architectural character.

Together with the exhibition houses a handsome catalog was produced, which extolled the virtues of the copper house in a twelve-point manifesto, answering the self-posed question "Why is the copper house the best?" Attention was drawn to its precision, being mechanically assembled; its hygienic qualities; its efficient thermal insulation, which made it economical to heat; its proof against fire, lightning, and earthquake; and the fact that it could be erected in 24 hours and internal partitions could be relocated. A price list for separate components—internal and external wall panels, glazed windows, doors, insulated roof decks—was included in the catalog. Wall panels were shown to have an outer facing of rectangular-patterned copper sheets and an inner lining of pressed steel, with a choice of six "tasteful patterns," some simulated brick or tile, others (the *Englisch* and *Japanisch* styles) with a delicate overall floral motif. These panels were modular, in 1-m increments up to a maximum of 4 m, and in two heights: 2.35 and 2.80 m. The wall unit, came complete from the factory, fully insulated with aluminium and asbestos and fitted where necessary with shuttered double-glazed casement windows (whose fanlights were remotely operated) or with doors. These large units were assembled simply and speedily by a crew of six workmen. All sanitary ware, electrical installations, and a fully-equipped kitchen were provided, with built-in cupboards available as optional extras. To round out the comprehensive specification, a choice of six internal colors was offered, ranging from coral red to pastel blue.

Nine houses were illustrated in the catalog. We see that the exhibition house in Berlin is representative of the architectural character of all models offered: conservative, with steep pitched roofs and romantic features such as arched porches (technically difficult to achieve in a metal-faced panel system) and flower boxes. To the prospective client a questionnaire was included, where he could tabulate the data relevant to his particular site and his preferences in relation to all the optional items: heating, built-in cupboards, sanitary equipment.

This was in May 1931. Early the next month we find *Bauwelt* approaching Gropius to prepare the material on the Hirsch houses for the *Bauwelt* catalog. To this request he responded, but publicity material and technical details were obviously supplied by Hirsch.[11] We have here what is evidently the beginning of a formal relationship between Gropius and Hirsch Kupfer. From June 1931 onwards[12] this relationship is expressed by the intensive and extensive activity undertaken by Gropius on Hirsch's behalf, activity that embraced not only architectural design but also technical research and development and market and sales promotion. This role was external to Hirsch Kupfer's organization but obviously much more than that of professional consultant. It was—and this is entirely consistent with the holistic philosophy of Walter Gropius—the provision of a total service. Gropius himself talked of having taken over the direction of the copper house division at Hirsch Kupfer.[13]

On 19 June 1931, having studied the model houses, and the documentation that had so far been presented to him, Gropius put forward his views in a wide-ranging evalua-

4.5
Hirsch Kupfer-und Messingwerke, internal finishes to copper houses, 1931

4.6
Hirsch Kupfer-und Messingwerke, window detail, copper houses (Grundman House, Haifa, 1933–34, by Deutsche Kupferhaus Gesellschaft).

4.7
Hirsch Kupfer-und Messingwerke, erection procedure, copper house, 1931

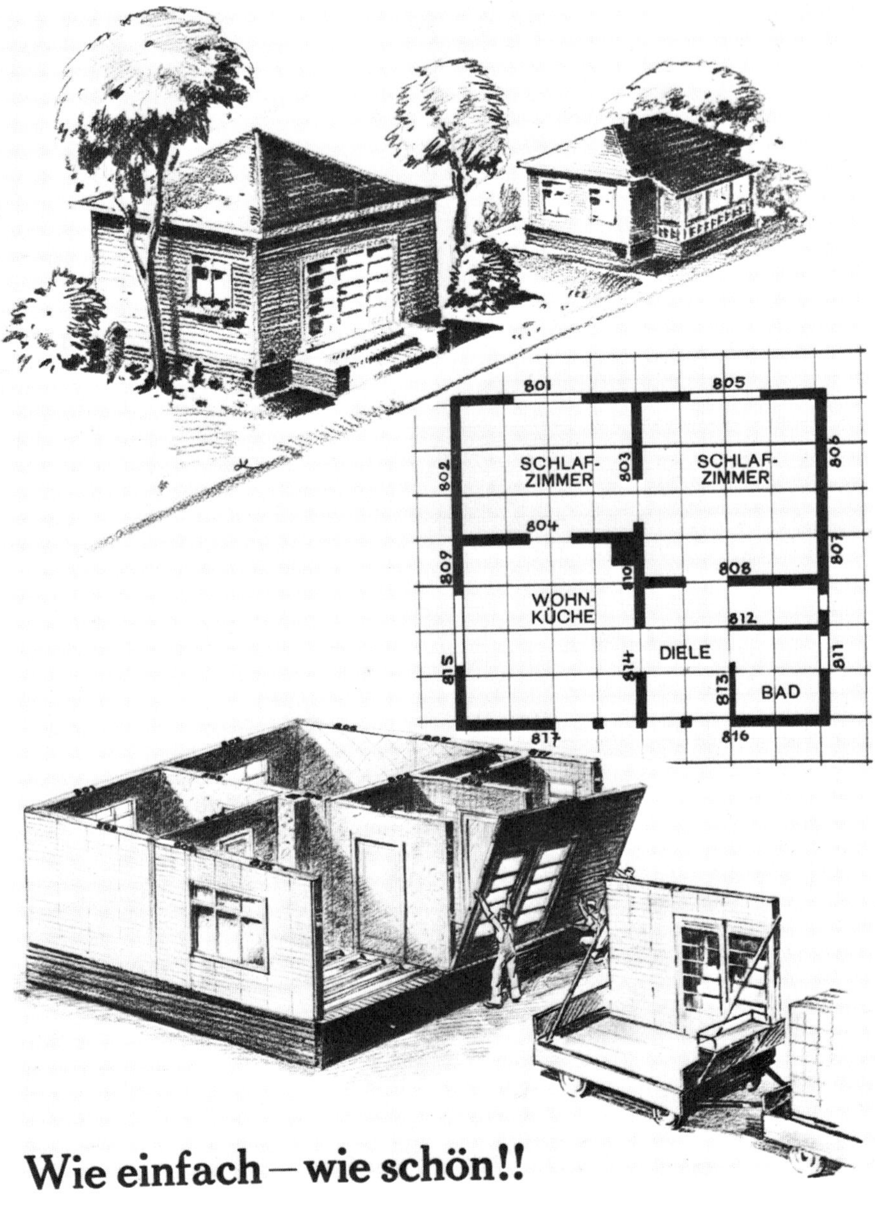

4.8
Hirsch Kupfer-und Messingwerke, "Kupfercastell" type, 1931

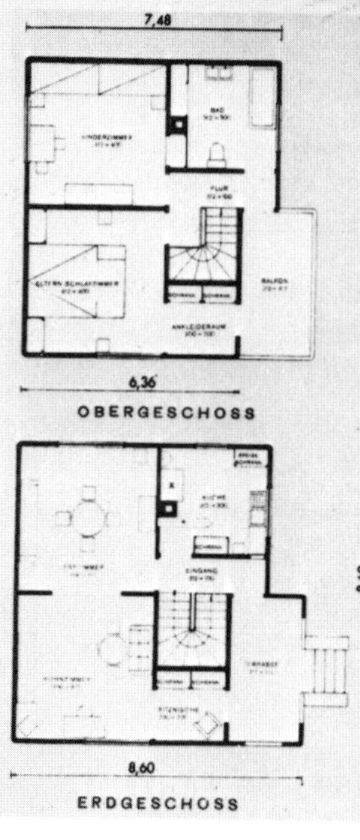

Haus „Kupfercastell"

Hier ist ein entzückendes zweistöckiges Eigenheim, das durch sein gefälliges Aeussere und seine gelungene Linienführung mit Recht das Ideal eines zweistöckigen Einfamilienhauses genannt wird. Interessant ist die Anlage der Eingangsterrasse, die nach oben hin in einen grossen Balkon endigt, der auch vom Elternschlafzimmer aus zu betreten ist. Die Eingangsterrasse, die eine enge Verbindung mit den Innenräumen herstellt, ist ein beliebter Aufenthaltsort, eine Lebensfreude für den Besitzer. Ueber die Terrasse gelangt man zu dem seitlich gelegenen Eingang. Hier eröffnen sich die unteren grossen Räume. Links von der Diele sieht man eine bequem begehbare Treppe, die zu den oberen Räumlichkeiten führt. Das Esszimmer bietet durch eine reizvolle Oeffnung der Zwischenwand einen wirkungsvollen Durchblick zu dem geräumigen Wohnzimmer mit seiner idyllischen Sitznische. Die Anordnung der Fenster ermöglicht eine rasche Durchlüftung der Räume. Die Küche ist mit allem Komfort ausgestattet, um die häusliche Arbeit so angenehm wie möglich zu gestalten: passend eingebaute Schränke, eingebautes Bügelbrett, ein gut gewählter Platz für Herd und Abwaschtisch sind die besonders auffallenden Vorzüge. In unmittelbarer Nähe des Kücheneingangs bietet sich, falls eine Unterkellerung gewählt wird, die Möglichkeit zur Anlage eines Kellereinganges, der seine Beleuchtung durch ein Fenster von der Terrasse empfängt. Von der im ersten Stock gelegenen Diele führt ein separater Eingang zu den beiden Schlafzimmern, dem grossen Bad und dem Balkon. Einen ganz besonderen Komfort bildet die Ankleidenische im grossen Schlafzimmer mit den beiden eingebauten Wäsche- und Kleiderschränken. Ein Austritt zu dem Balkon ist auch hier vorgesehen. In den Schlafzimmern wie im Bad sind unsere eingebauten Schränke eine sehr willkommene Einrichtung. Der Bodenraum ist mit einer Umklappleiter von dem oberen Flur aus bequem zu erreichen.

BLICK IN DAS WOHNZIMMER

tion of the copper houses.[14] This paper is impressive both for its breadth of scope and for its precise grasp of detail. On the one hand, Gropius deals with broad questions of housing policy and dwelling preferences; the conflict between industrialized building and the protective nature of conventional building workers' organizations; the need for standardizing parts while maintaining the variability of the whole building. On the other hand, he produces a six-page summary of detailed criticism of technical aspects of the design, including structural safety, construction details, condensation and waterproofing, electrical and plumbing installations, maintenance and cleaning problems, even the noise problems of heavy rain or hail on the copper sheets.[15] He considers the advantages of fixed prices and guaranteed delivery dates, and the legal obligations deriving from these which will fall on the manufacturing company. He proposes to separate the manufacturing from the merchandising functions and advocates for the latter the establishment of a separate marketing and finance company. We may at this point comment on the continuity in his thinking here, in relation to his pioneer manifesto: "Program for the establishment of a company for the provision of housing on aesthetically consistent principles," which he had presented to Rathenau of AEG in 1910. Gropius now proposes a plan of action:

I recommend that you stop advertising the houses for the time being and conclude only those sales that arise out of the exhibitions in Berlin and Paris. Your gentlemen dealing with sales assume that with the conclusion of the exhibitions, approximately 20–30 sales would have been made to the month of August, so that the factory would be rather busy in the immediate future. As a second stage, I propose the development of the types, taking into consideration the design and technical faults that have been complained of. This would be done during the winter months, so that, following these preparations, active advertising may be renewed in the spring. The third stage should be started at the same time, namely the development of future houses, technically more improved . . . .

Finally, Gropius recommended that a limited series of standard types be developed, capable of satisfying a wide variety of demands. We have no documentary evidence that Hirsch Kupfer formally accepted this program, but the subsequent facts speak for themselves. For the immediate future the direction indicated by Walter Gropius becomes the policy of Hirsch Kupfer.

Improving the Design

Hirsch and Gropius were united in their desire to refine the copper house, eradicate the technical problems, and enhance the performance in every respect. They therefore entered into a long and serious process, whereby the prototypes of the Berlin exhibition were probed, tested, evaluated, and improved. These tests took three forms: the examination of elements of the building system, and systems of manufacture and assembly, mainly in Hirsch's own workshops in the Copper House Department at Finow, near Eberswalde; the seeking out of expert opinion and the evaluation of materials, processes, and environmental performance by key German research institutes in such fields as acoustics, thermal insulation, sanitation, heating, safety in terms of fire and lightning, and so on;[16] and the building at Finow of an experimental *Siedlung*, originally of six houses, for the testing of materials and techniques under field conditions, over protracted periods of several months. When shortcomings were revealed, or when more efficient methods thought of, alterations were effected, comparative analyses made, and the consequent improvements built into the design and construction processes. The policy, both in relation to the original prototypes and later to the new types developed by Gropius himself, was one of continuous evolution and improvement rather than radical change. At all times the implications of changes suggested for aesthetic or technical reasons upon the cost, or upon the processes of industrial manufacture in the factory, were carefully balanced. And a constant dialogue was maintained by the technicians of the Hirsch plant and the designers in Gropius' office in that sort of creative interaction between architect and industry presupposed in the foundation of the Deutscher Werkbund and adumbrated in the educational programs of the Bauhaus.

Although all aspects of the design and construction of the copper houses were subjected to this close and ongoing scrutiny—the foundations, external staircases and railings, waterproofing details, roof construction and covering details, electrical and sanitary installations, fireplaces, stoves,[17] and central heating (even to a comparative analysis of the cost of alternative fuels)—the most serious attention was undoubtedly given to the wall panels, which were the system's most unique and characteristic feature. Tests were mounted and discussions held on the copper sheets, which constituted the external face of the panels, examining their thickness and metallurgical qualities, the rigidity and stability of their pressings, the problems of patina. Similarly, the inner facings of pressed steel sheets were closely studied, and alternatives, particularly of aluminium,[18] evaluated and eventually adopted. Both aluminium and steel[19] were investigated by Gropius as possible alternatives to the copper facing. And though we may assume that Hirsch was not too happy at this attempt to bypass copper, their raison d'être, they at no times stopped these open-minded queries. The construction of the panels and their connections formed the basis for much debate, and some controversy. Gropius and Hirsch examined, in drawings and model form, the possibilities of an ingenious jointing system that eliminated the need for the U-section connectors and fixed the panels directly together by means of spring-loaded bolts. However, despite advantages, including the elimination of unsightly cover strips, it was rejected as involving many new problems. It could mean new tooling in the workshops, more complicated fixing patterns, a deviation from the standard module, and an upset in packaging arrangements. The question of modularity and standardization was a problem to which Gropius in particular was especially sensitive.[20]

Most obdurate of all problems, however, was the question of thermal insulation. This had been at the core of Förster and Krafft's original patent for the wall panels, and endless attempts were made to upgrade the performance over a period of six months. Gropius, out of his own experience at the Weissenhofsiedlung[21] preferred metal foil and supported the introduction of the Dyckerhoff system, using aluminium, subject to proper testing. An arrangement was entered into with Dyckerhoff, who was paid licence fees of RM 6,000 for a period of about six months, expiring on 31 December 1931.[22] Tests were done at Hirsch's, and also by independent sources at Munich; alternative designs were prepared, and comparative evaluations made, some in the presence of Förster and representatives of Dyckerhoff. In the final analysis such good results were obtained that Gropius could cite the high thermal value to weight ratio of the walls as his reason for considering the system the best dry-assembled method he had ever come across. The ingenuity of Förster, the rationality of Gropius, and the technological integrity of Hirsch Kupfer combined to make it so.

Together with this process of continuous improvement, Gropius turned his attention increasingly to the development of new house types. The original designs, as we have seen, were open to criticism on both aesthetic and

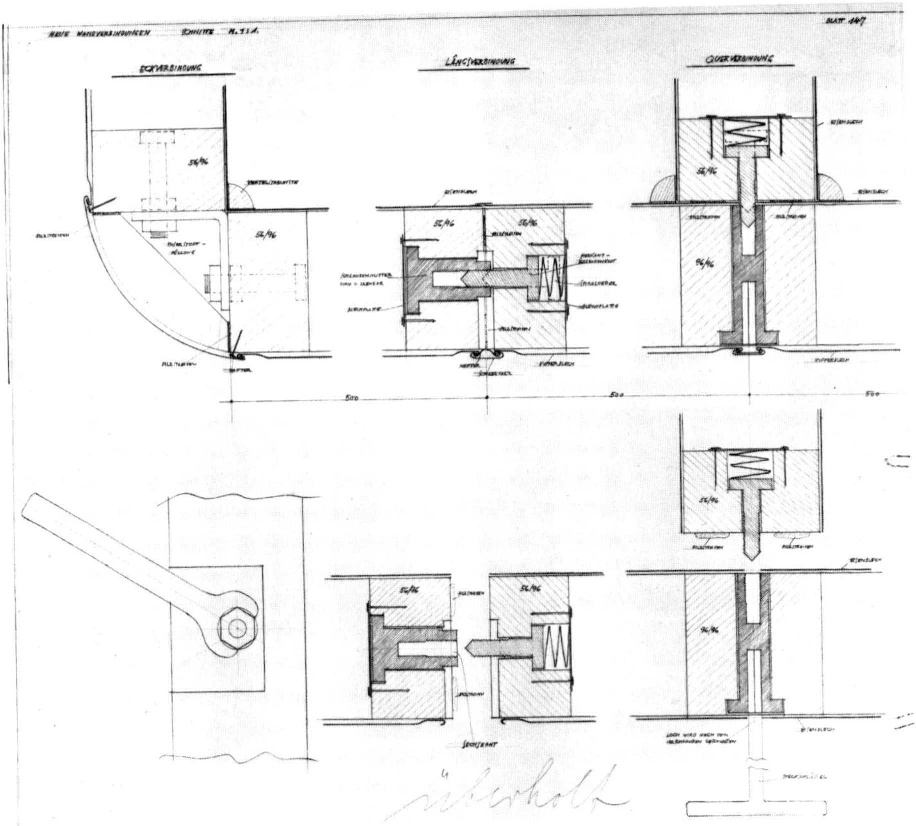

4.9
Walter Gropius, proposed new connection system for Hirsch panels, 1931–32

technical grounds. In a letter to Siegfried Hirsch,[23] Gropius maintained his positive view of the copper houses, seeing no inherent problem with respect to both the plan and the outer appearance that could not be overcome by careful design. The decision of the inventors of the copper houses to show them to the public in an "old-fashioned style" was, he felt, mistaken, for the reactions of the public to the houses at the exhibitions of Paris and Berlin had been negative. The public Gropius had in mind was an elite one, "the more refined public, architects, and experts in taste" who rejected "the imitative character of the houses." However, he went on, these difficulties "could be overcome by improvement of the types and the form of the houses, by reworking and improvement."

In July 1931 Gropius began this process. The original list of types offered for sale by Hirsch Kupfer consisted of a basic nine ranging from the most expensive of the list, "Kupfercastell," of 100 sq m, costing RM 10,900, to the cheapest, "Eigen Scholle," of 56.6 sq m, at RM 6,300. To this list was added an unnamed type K, a modest 36.9-sq-m dwelling costing only RM 4,200, and type R, a 167-sq-m seven-roomed villa, at RM 13,600.[24]

Gropius set to work upon the K type. By the end of July the design had stabilized, as a basic (type K) core house of 37 sq m,[25] containing a minimal three rooms, kitchen alcove, and WC, which could be expanded by the addition of a further two rooms to make a new variant (type K1) of 62 sq m. Another related pair of designs, K0 and K2, represented a slightly modified version, with separate kitchen, of both the core and the expanded houses. A type L house was also developed, with both single- and double- story variants, but apparently this type was abandoned in favor of type K.[26] Gropius then went on to develop the M series, whose basic unit of about 49 sq m (type M) could be repeated to form a double-story house roomed dwellings. The large type R house was also further developed, as well as two variants of type D, the original ''Juwel'' house of the Hirsch range.

By mid-July preliminary sketches were complete and some plans already incorporated by Hirsch Kupfer into their catalog. A week later prices had been calculated for five variants; by the end of the month many details had been established, including a new roof system, the question of heating, quotations for ironwork for the outside stair, and a precise schedule of areas and prices.[27] During the next two months an intensive campaign of design and development took place. New drawings continued to be prepared in great numbers. By 19 November 1931 Gropius' office had prepared some 130 sheets of drawings for what was called the ''Gropius-Typen,'' that is, types K and M, about half of which were superseded in the continuous evolutionary process of refinement. Carefully drawn sketch plans were prepared of all variants, showing plans and elevations. Rendered perspectives showed attractive little houses of restrained modernity, all with pitched roofs, some hipped, others with gables. There is, in fact, a striking resemblance to the architecture of the *Montagehäuser* houses, with their simple fenestration and wire-meshed balconies, which Gropius had designed in 1929. In addition to these presentation drawings all types were studied from the point of view of solar penetration, according to various orientations. These drawings were carried out in Gropius' atelier.[28] After much development selected versions of each type were then detailed in working drawings and detailed studies, and well over 100 sheets of such drawings were prepared in the Hirsch Kupfer drawing office (Abteilung Kupferhausbau). This ongoing, evolutionary design process is what Gropius had in mind when he advised Hirsch on the necessity for reworking and improving the planning and appearance of the copper houses. This process included the building of a prototype of the K type on the Wolfwinkelerstrasse at Finow, both as a show house and as an object for experiment; construction of this house was still continuing when Gropius visited the site early in January 1932. The lessons from the prototype were then fed back into the revised standard detailed drawings of the models meant for mass production (Gropius talks at an early stage of an output of ten houses per month[29]), and into revisions in the tooling and flow chart of the factory itself. As notes made by Gropius after a meeting with Hirsch Kupfer at Eberswalde, on 21 October 1931, indicate, the final drawings for types K and M depended on the results of the Wolfwinkelerstrasse experiment, talks with technicians about manufacturing problems, and decisions about assembly flow lines. These notes included a list of problems needing attention, some for technical reasons, others (relating to the internal and external skins of the wall panels) for aesthetic reasons. Gropius concluded, cautiously and characteristically: ''All improvements must be done successively, after testing, because the process in the factory cannot easily be altered.'' As much as he wanted to see the ''gradual alteration of dwelling types in both their technical and formal aspects, considering the customs and tastes of other countries,'' it would be best, he felt, to wait and see if sufficient houses were sold to make this worthwhile.[30]

This considerable effort in the copper house venture, together with the houses then being considered for the next Berlin exhibition (of which we shall have more to say later), compelled Gropius to give serious thought to the organization of his own office and the division of responsibilities between himself and Hirsch Kupfer. This he formalized in a memorandum early in January 1932.[31] As he saw it, the Berlin office of Hirsch Kupfer should be responsible for calculating all quantities for the foundations

4.10
Walter Gropius, type K1 house for Hirsch Kupfer, 1931

4.11
Walter Gropius, type M2 house for Hirsch Kupfer, 1931

and superstructure, structural calculations, cost estimates and flow charts, formal presentations to building inspectors, negotiations on materials, the preparation of detailed assembly manuals, and the preparation of sales catalogs. Gropius of course would provide the drawings for the catalogs, as well as the preparatory drawings for all estimates of costs and quantities. The main work of his office would naturally be the preparation of all necessary drawings. He suggested setting up a special atelier, under the direction of Herr Dustmann, with six assistants: Goetz to work generally on details; Fieger on sketches for types IV (''Juwel'' or type D in the original list) and V (type R), as well as detailed studies; Luderer, concentrating on type M, with all its variants; and Hadda, Gumberz, and Okamura principally engaged on the K types. Gropius rather hoped that Hirsch Kupfer might be able to spare a draftsman to help out in his office, and he thought it desirable that there be coordination between his office and the man from Hirsch Kupfer responsible for production design. One can sense in Gropius a deep concern over the responsibility of documenting designs for industrial production. In craft buildings, he pointed out, measurements and adjustments could always be made on the site but production in factories demanded absolutely accurate documentation, as an error was reproduced not once but innumerable times.

Gropius, then, was heavily involved in the ongoing design process for the copper houses as well as in their technical evolution and perfection. But this was by no means the sum total of his commitment to Hirsch Kupfer and the copper house project.

The Selling of the Copper Houses

Before the war, when Gropius had worked in the office of Peter Behrens, he had been aware of the full range of services—advertising, industrial design, architecture—which his mentor had provided for his client, the AEG. Now Gropius was similarly called on by Hirsch for many services beyond the purely professional, and he became deeply involved with the promotional aspects of the copper house business. He maintained contact with the daily and professional press and was consulted by Hirsch on poster design and other advertising material.[32] He took a central role in the design of their proposed new catalog. He prepared drawings, had models constructed and photographed, designed the layouts, and negotiated with the printers over such details as paper types and cost. His interest in graphic design went back of course to his Bauhaus days, and it was perhaps natural for a universal designer such as Gropius to be engaged in these matters. But his involvement went much deeper, into the financial and organizational aspects of the business, and in the active promotion of the copper houses as products for the home and foreign markets. It must be stressed that in this he played as much the role of entrepreneur as that of designer: indeed, it would have been difficult to separate the two aspects.

We do not have any accurate measure of the extent of the trade in copper houses at this time, for although we have some of Gropius' records, none of Hirsch Kupfer's appear to have survived. During the first half-year or so of Gropius' connection with Hirsch Kupfer, that is, until the end of 1931, we have firm knowledge of 30 houses actually produced by the copper house division. Of these, 13 were for their own use—6 for the exhibitions at Paris and Berlin, 6 for the model settlement at Finow, and a further test model built there according to Gropius' K type. In addition, from various sources,[33] we know of 17 for one family (M1, M3) or two families (M2), with an external stair serving the upper of two superimposed three-

4.12
Walter Gropius, drawing of type M1 copper house prepared for Hirsch Kupfer catalog, 1931

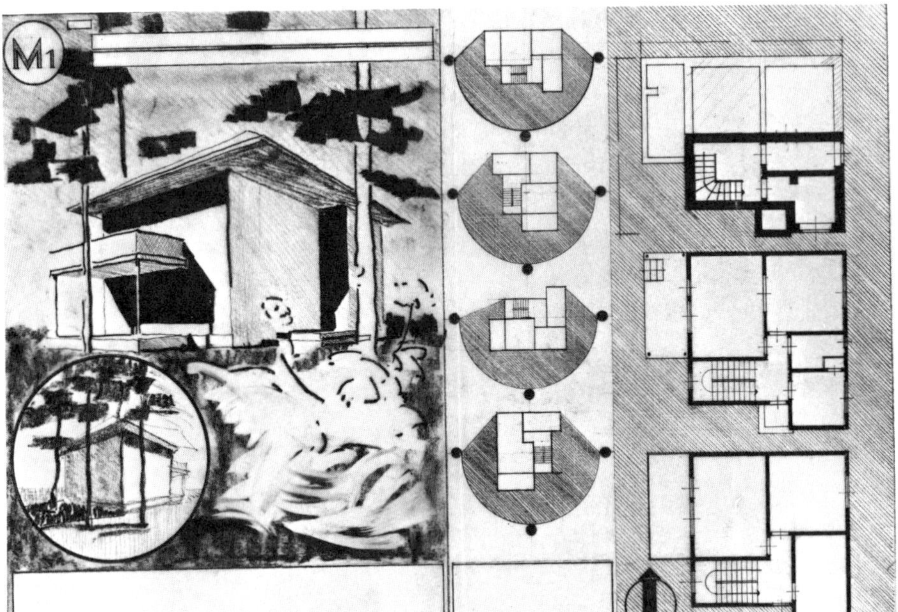

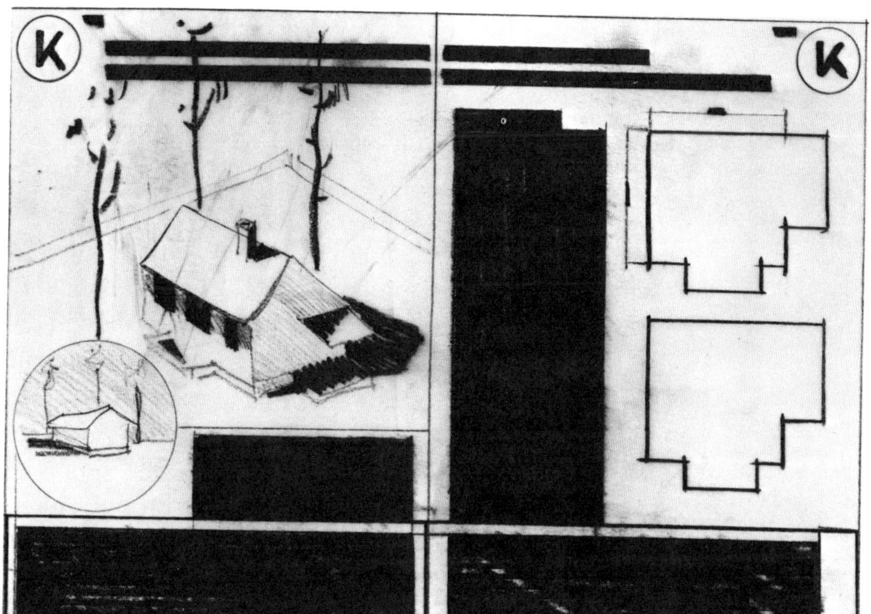

4.13
Walter Gropius, dummy layout of catalog page, 1931

others built for private customers and erected in Germany, mostly in the Berlin area. Of these, the most popular was the large "Kupfercastell" model, with some interest also in the "Juwel" and "Fruhlingstraum" types. These were types, according to a letter from Hirsch Kupfer to Gropius, which, because of their functional design, were in use all the time at the copper house division.[34] Gropius, thinking back,[35] thought that some 53 copper houses had been erected, presumably during the period of his active association with Hirsch Kupfer.

In addition to these certain examples, we know of many more potential cases, few, if any, of which could have materialized. In Gropius' own records we have notes of many deals being negotiated, some for considerable quantities. Shortly after the commencement of the Berlin Building Exhibition (in May 1931) we have notes (exchanged between Gropius and the sales department of Hirsch Kupfer) of discussions with Scher and Littauer, owners of a tract of some 2.5 million square meters near Glienicke, who were interested in starting a finance company to deal with the copper houses; with Lindemann, interested in erecting 8 six-roomed houses in Grunewalde; with the magistrate of the town of Cottbus, for 12 six-roomed houses built under a guarantee from the town; for the Heika (Heiga?) project in Berlin, which spoke of a large number of different types in the RM 5,000 price range, with 50 to 100 houses of each type; with Israel and Schulzendorf of Eichwalde, interested in similar houses for their large subdivision of land; with Professor Lazarus, who had on 15 June 1931 requested a special price for an offer to buy 33 houses; and for the erection of a small-house *Siedlung* in Berlin, consisting of several hundred low-cost houses. These numerous inquiries were very encouraging. Nor were they a mere flash in the pan, because as late as April 1932 Gropius—not ever given to easy optimism or to exaggeration— could write of the relatively large number of inquiries (several hundreds per month) which indicated the interest of the buying public in the copper houses. This interest moreover was not confined to Germany alone. "Numerous foreign connections have shown a lively interest abroad (Belgium, America, France, Rumania, South Africa)."[36]

The international connection of Hirsch Kupfer and Gropius' involvement in the major effort to export the copper houses make interesting reading. We may assume that from the very outset, with their decision to exhibit at the Paris Colonial Exhibition in 1931, Hirsch Kupfer looked abroad for a large potential market. Even before houses were actually produced, in fact even before proper publicity material was ready, at the end of 1930, Thiess & Co. of Hamburg, who represented Hirsch in Latin America, had promoted the copper house idea "as a novelty" in Buenos Aires, evoking the interest of experts and estate agents. It had been decided to leave the matter in abeyance until May or June 1931, when Hirsch hoped to be ready with the necessary material.[37] In July 1931 Gropius, apparently unaware of the previous activity, suggested to Hirsch that a Herr Moller, who was going to Buenos Aires to open a branch office for Gropius, might handle all the copper house business in South America.[38] This suggestion stirred up something of a hornets' nest, both in Thiess & Co. and in the sales department of Hirsch Kupfer, but an exchange of firm letters soon settled matters and left the South American business in Thiess's capable and experienced hands. What materialized from this transaction, in the form of actual exports of copper houses, we unfortunately do not know, and there is this same tantalizing gap in our records in nearly all the following incidents. They are important, and are here recorded, if not for the tangible results (which presumably were negligible), then for the light they throw on the interest overseas in the copper houses and the strenuous efforts made by Hirsch Kupfer and by Gropius to exploit this interest and promote further the export of their prefabricated products.

In this search for overseas markets we have some incidental cases: a cryptic reference to houses for the Princess de Ligne, for Sir Robert Williams in Africa, and a "Juwel" type house for a Mr. Hoboken in Holochem (Hollochne?), in Belgium, which was under construction in July 1931.[39] The Belgian connection is significant because the giant copper firm of Union Miniere of Haut Katanga, with headquarters in Brussels, were shareholders[40] in Hirsch Kupfer-und Messingwerke. There is some discussion between the partners, examining the possibility of exporting copper roofs in component form (of the type made for the copper houses) and the copper house itself. This latter was examined in some detail by the Brussels firm, which had established its own Copper House Department, and a translation of their findings, which including criticism of some technical details and the ventilation of the houses, was kept by Gropius in his files.[41] Another associate, Anaconda Copper Mining Co. of New York, also showed interest in these new products. They ordered, probably in June 1931, two models of the "Fruhlingstraum" house and one of the two-story "Kupfercastell." By August, letters and cables were exchanged, because of the nondelivery of the necessary specifications and other documents and—more seriously—because Anaconda, after submitting details of the houses to its experts, concluded that changes were necessary to adapt them to the American market. These changes centered on the need to increase the window area, to improve the interior design to meet the high standards demanded by the American housewife, and to alter the electrical installation to meet the requirements of the National Electric Code. New York, Los Angeles, and Houston were considered the likely locations of the three houses, and Anaconda asked Hirsch Kupfer to supply detailed shipping information.[42] From this information, which had already been meticulously worked out, we learn that the "Kupfer-castell" house was shipped in 34 packages, weighed 15,513 kg, and occupied 69.68 cu m of shipping space, whereas the "Fruhlingstraum" was packed in 31 packages, weighed 15,901 kg, and had a shipping volume of 73.48 cu m.[43]

In August 1931 a Mr. Malletke, a business associate of Gropius' (whose interest in the copper houses had come about through the intervention of Gropius' kinsman, Alfred Gropius) brought Gropius into contact with the Budapest office of The Union Guarantee and Industrial Trust Limited, the foreign division of a British financial company. To them Gropius wrote, as head of the Copper House Department of Hirsch Kupfer, sending the old catalog, the folder containing details of his new types, and many prints and photographs. He drew attention to the economy and high quality of the copper houses, their light weight and ease of transportation, and the short and guaranteed delivery date (two to three months, including all the time necessary for government formalities and financing arrangements). An immediate and encouraging reply stated: "We are so much interested in the project that the undersigned will come to Berlin either this or next week and will talk to you about the matter." Triangular negotiations between Berlin, Budapest, and London went on for some time, but the final outcome is not recorded in the Gropius files, only a highly optimistic note from Union Guarantee suggesting that all the details had successfully been negotiated between Gropius and their Mr. Tilley.[44]

The last of these international episodes in which Gropius was heavily involved was left equally unresolved, but it throws fascinating light on the relations between the Soviet Union and its well-wishers in the west, before the heavy hand of Stalin severs the links. On 2 August 1931 Gropius reported an interesting encounter to the sales department of Hirsch Kupfer:

On Saturday the president of the section of Foreign Specialists of RKI in Moscow, Herr Ing. S. J. Rutgers and Dr. F. Frankl, Moscow, came to consult me on houses that can be quickly manufactured as dwellings for foreign specialists in Russia. I gave them detailed information on the construction of the copper houses and showed them all the types. They were greatly interested, and asked for an immediate proposal, which I shall describe more exactly. The idea of these people is the following. Because of the difficult foreign currency situation in Russia they want to ask engineers from outside Russia to bring with them their own houses, which will be paid for in German currency, for which they will be refunded in rubles. Because of the urgent need for such dwellings a guarantee by the state of the payment in cash in foreign currency may be possible. These gentlemen asked me for an immediate quotation for types K, K1, M2, M1, and they want a proposal for one or for fifty units, sent to Stettin or to the Polish-Russian border.

Gropius then goes on to discuss various detailed changes that would be necessary, especially because of the great cold in Russia. Although the Russians themselves would be responsible for the assembly of the houses, they wanted data "on the number of assemblyworkers and hours of erection, according to our calculations."[45] As these proposals were being prepared by Hirsch Kupfer, and a proposal was being considered to send to Munich for tests of the insulation for a possible temperature of −25°C, Gropius pressed his case directly in Moscow through the mediation of his old friend Ernst May. The former Frankfurt architect had gone to the Soviet Union as the head of a task force of architects and planners,[46] some of the many "newly engaged specialists from America and Germany" to whom Gropius referred in his letter. As he explained the situation to May: "I think that the copper houses of my newly developed project are especially suitable, after we have eradicated some small faults, which you too have remarked on. Since you left I have had several tests made in various institutes, with very good results, and because of the extraordinary insulation capacity of the wall I do not hesitate to recommend these houses for Russia. . . . I ask you to support this matter, so that a shipment of our first series, perhaps a trial shipment, may be realized."[47]

None of these international ventures seemed to come to maturity, despite Gropius' active participation in their promotion. It is true that economic conditions in the world as a whole were not, in these early years of the Great Depression, conducive to grand experiments and new ventures. In fact Gropius' enthusiasm and ambitions somewhat alarmed the more conservative heads at Hirsch Kupfer. A memorandum of July 1931[48] notes the hopes of Gropius and Malletke, supported by René Schwartz of the Hirsch Kupfer directorate, to sell houses on a big scale during that year. They believed their good connections, and the resultant financial means, would enable them to carry out the project. Hirsch Kupfer, however, favored restraint, suggesting that no massive sales campaign be undertaken until the winter tests had been successfully negotiated. In preparing carefully for launching the house on the market, many points had to be considered, including discussions with companies responsible for rural and municipal housing estates (of which there were more than a hundred in Berlin alone), negotiations with estate agents specializing in housing, negotiations with the authorities, assurances of credits and proper financial agreements entered into, advertisement abroad, and organization for selling. It was considered that Hirsch

Kupfer would not really be ready to commence selling houses until April 1932 and that considerable money had to be invested in the effort before then. It was possible, concluded the memorandum, that a trading and finance company, on the American model, should be set up for this purpose. Schwartz himself looked into the question of the financial arrangements with savings banks. Although attention was drawn to the possibility of difficulties in raising mortgages, or arranging insurance, for the copper houses, "their being so easily transportable,"[49] this did not prove to be an intractable problem. Mortgages of 60 to 70 percent of the cost were attainable,[50] and insurance against fire and burglary was available at the same rates as for the brick houses.[51]

The Wachsende Haus Competition

In Gropius' account of the advantages of his prefabricated copper houses, we find three recurrent themes, all indicating the flexibility and dynamic nature of the system. These themes are mobility, or the ease of transportation and adaptation to various locations and climes; adaptability, or the capacity to generate many house types and variations, through the interplay of standardized components;[52] and growth, or the expandability of the house, horizontally through the addition of further rooms, or vertically through the addition of another floor. We have already seen how the question of flexibility in the design of the dwelling was a significant area of innovation at this time. The question of staged growth, from initial core to expanded dwelling, had by 1931 become a topic of absorbing interest to many architects in Germany.[53] This interest was now to be focused on an event that gained considerable professional and public attention.

In response to the government program for encouraging the development of the *Stadtrandsiedlung*, or peripheral urban settlement,[54] a competition for ''Das Wachsendes Haus'' was announced in Berlin in October 1931, with designs to be submitted by the end of the year. Martin Wagner, *Stadtbaurat* for the City of Berlin (whose Bureau for Fairs and Exhibitions was the principal sponsor of the competition) played a major role in its administration. As it was intended to incorporate the results of the competition in the Berlin Summer Exhibition of 1932—''Dwelling for All'' (*Sonne Luft und Haus für Alle*), as it was somewhat optimistically to be called—Wagner expanded its scope by setting up a working group (*Arbeitsgemeinschaft*) of twelve eminent architects to produce additional designs.[55] This group, which included Gropius, Otto Bartning, Mendelsohn, Max and Bruno Taut, and Poelzig, were set the same task as the competition entrants: to design an expandable house, with a core of 25 sq m and at a cost not exceeding RM 2,500. The increments of growth should be small, it was specified, and expansion should be accomplished without disturbing the ''livability'' of the house to the occupants. It was hypothesized that industrialization would be used to reduce costs—but the ultimate test of this, it was realistically concluded, would be when the houses were actually built. Standardization was desirable, but the individual needs of each owner should not be sacrificed. In all, it was an eminently sensible and humane brief, and for its day a remarkably forward-looking one. It brought together two main lines of development: dwelling flexibility and industrialized building systems.

The architects' response to this challenge was encouraging. Despite the short time allowed (six weeks from announcement to judgment, six months from judgment to the opening of the exhibition) over 1,000 competition entries were received from architects in Germany. We must remember, however, that the architectural profession was desperate for employment at this time, with massive unemployment in the building industry generally and over 90 percent of architects without work.[56]

When the Growing House exhibition opened at the Berlin Exhibition Grounds in May 1932, there was eventually to be seen not only a comprehensive display of drawings and models but complete houses at full size designed by 23 architects, including submissions by the working group and prize-winning schemes from the competition.[57] A wide-ranging debate ensued in the architectural journals on many aspects of the problem, including the principle of growth; the economic questions raised by deferred investment, and the effect on mortgage policies; the detailed comparative analysis of alternate building systems especially in terms of performance; and the question of self-help.[58] The extensive publicity deriving from the exhibition, the comprehensive catalog, the coverage in the professional press, and Wagner's subsequent book helped to generate a climate of interest.[59] Further work was done in the field, and a similar Growing House exhibition was mounted in Vienna.[60]

In the Wachsende Haus exhibition in Berlin in 1932, many of the demonstration houses used industrially produced systems. Some were based on modular structural systems, with lightweight steel or timber frames. Others utilized panel systems, wooden framed with wood, asbestos or metal cladding as infill panels to the structural skeleton or as independent self-supporting structural wall elements. The ingenuity and sophistication of these systems is impressive. At least three architects—Alfred Gellhorn,[61] Erich Mendelsohn, and Rambald von Steinbuchel-Rheinwall used the by now well-known steel system of Böhler-Stahlbau, of Berlin, to which we earlier referred. The Taut brothers, Max and Bruno, each contributed a house erected by the contractor Philipp Holzmann, the builder of some of the earliest steel-framed apartment buildings in Germany, notably for Mebes and Emmerich at Berlin-Britz, in 1926.[62] At the Berlin exhibition Holzmann was again to work with Mebes and Emmmerich, using the Dawa[63] system of lightweight concrete as against the "Müller-Holzmannschen Methode" of steel framing and concrete panels for the Tauts, or a more radical arcuated system for Hans Poelzig. Bartning's "Werfthaus System Bartning" is of especial interest for its ingenious use of triangular steel studs to which the modular copper-coated steel wall panels are attached structurally but separated by a rubber buffer strip—a sophisticated technology deriving from Bartning's collaboration on the project with "the engineers of an airplane factory"—at least according to Wagner.[64] Wagner himself was particularly impressed by two schemes: those of Bartning and Walter Gropius. Despite Gropius' long-held interest in steel-based systems, including the *Montagehäuser* houses, it was obvious, and inevitable, that for the Wachsende Haus exhibition he would work in collaboration with Hirsch Kupfer- und Messingwerke. It was in a sense the logical climax to the intensive months of design and development they had undergone in partnership.

The first specific reference in the Gropius records to the "Wachsende Haus" is found in a letter of 5 March 1932 when, in discussing the possible replacement of the inner steel wall linings with aluminium sheet, he suggested to the aluminium company concerned "cooperation on both of the experimental houses which I will be exhibiting with Hirsch Kupfer at the Building Exhibition."[65] Now, although this letter shows that the specifications of the exhibition houses were still far from settled in March, we must assume that the design of these houses commenced much earlier, perhaps in December 1931, when Wagner included Gropius in his *Arbeitsgemeinschaft*. These sketch plans in any event were ready by January 1932, when the work of the group was first published in the profes-

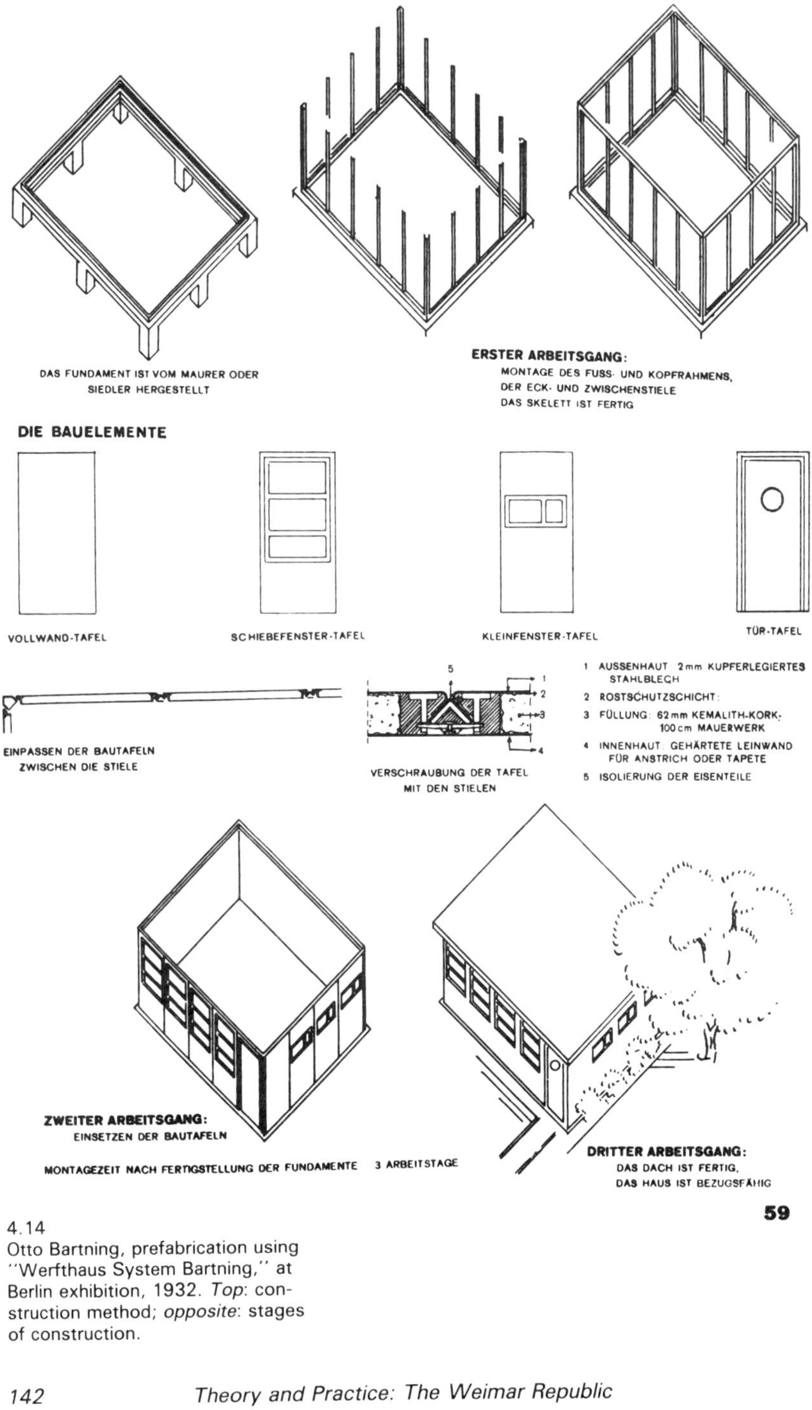

4.14
Otto Bartning, prefabrication using "Werfthaus System Bartning," at Berlin exhibition, 1932. *Top*: construction method; *opposite*: stages of construction.

9.00 UHR FRÜH

11.00 UHR VORMITTAG

2.50 UHR NACHMITTAG

4.00 UHR NACHMITTAG

sional press.[66] Although none of the drawings prepared in the Gropius atelier for this project are dated, several prints bear the approval stamp of the Hochbaudeputation Berlin (signed: Klieberswirth) and dated 15 December 1931. The design itself, basically an L-shaped plan, unusual for Gropius,[67] underwent many permutations—some remarkably inept—before the final solution was evolved. Over 20 sheets of working drawings and details were then prepared, mostly during March 1932.[68]

Time was now running out, if the exhibition deadlines were to be met. The soul of efficiency and dependability as always, Gropius prepared a detailed work schedule covering all subcontractors with a timetable extending from 18 April to 10 May 1932, a minimal period indeed. The formal contract for the erection of the houses was signed late, toward the end of April 1932, which did not give much time to have everything ready for the May opening. Hirsch Kupfer, the official exhibitors, wrote to Gropius, defining his responsibilities: "We have signed the application for the erection of two copper houses at the Building Fair, and are therefore obliged to erect these houses. We instruct you to carry out the erection of these houses according to the agreed drawings, and put at your disposal a budget of RM 8,000. . . . You have taken upon yourself to handle the deliveries in our name, and to cover the bills out of this budget. It is also your task to negotiate with potential agents. Naturally you will see that our interests will be served. . . ."[69] By mid-May this task was honorably fulfilled. The site layout was planned, the foundations were built, and from the factory arrived the trucks bearing the components which were duly erected by a trained team of men. Two handsome houses stood in the Berlin exhibition grounds, complete in every detail, and Gropius received a well-earned note of thanks from Hirsch Kupfer for the trouble he had taken on their behalf.[70]

4.15
Walter Gropius, Growing House using Hirsch system, transportation of components, Berlin exhibition, 1932

4.16
Walter Gropius, Growing House using Hirsch system, showing erection process, Berlin exhibition, 1932

Of all the houses shown at the Growing House exhibition of 1932, there is no doubt that those of Gropius were the most widely acclaimed and publicized.[71] Not only did his by now formidable reputation lend credibility to his experiments, but the technical quality and fine appearance of the houses won for them considerable approval. They were of course based in principle on the original Hirsch system and utilized the basic patented panels and jointing methods of Förster and Krafft. It is perhaps necessary to stress this dependency on the basic invention of Förster and Krafft because, where the original patent is mentioned as in Wagner's account of the exhibition, there is a tendency to treat it as a formality and not as a matter of substance.[72] Gropius, who had put so much effort into developing the project, may be forgiven for occasionally neglecting to give the originators of the system due credit: it is an overstatement of his role to claim, as he did in a lecture, that "a big metallurgical firm in Berlin employed my patent and drawings for the insulation of walls for manufacturing copper houses on an endless chain."[73] This does justice neither to Förster and Krafft nor to Hirsch. Gropius' contribution lay, not in initially developing the system, but in refining it technically and aesthetically. He planned the interior on efficient, functional lines in the best Bauhaus tradition. He replaced the inner wall lining of steel with aluminium and improved the thermal, acoustic, and other properties (corrosion, hygiene, fire resistance, etc.) of the system after the most rigid testing by the most competent scientific authorities of the day. Moreover he simplified the external appearance, used a horizontally ribbed corrugated copper sheet instead of the rather clumsy original, replaced the pitched roof by a flat roof (mandatory, it would seem, for the International Style), improved the proportions, and transformed the rather mundane and traditional-looking Hirsch prototype into a prefabricated house of some elegance, which clearly expressed its modular nature. For Gropius it was a matter of no small satisfaction at last to be able to put into practice the doctrine of prefabrication which he had so consistently advocated since 1910, that doctrine which sought to unite the advantages of maximum standardization with the desired goal of variability.[74] However, the degree to which that goal was achieved in the Hirsch system is limited. There is inherent flexibility in the system, with a freedom to locate inner partitions freely, certainly at the planning stage, but with much more difficulty after construction. The possibility of planned growth is dramatically exhibited in the Berlin exhibition houses. Finally, there is in the system the inherent potential to generate many different house types, as the large number of types and variants developed in Gropius' studio demonstrate. Yet despite this, there is in Hirsch's approach, and in that demonstrated in Gropius' own design and development for Hirsch, a concentration on the house as a total product, and little if any evidence of a desire to develop a set of standardized components, as a language of design, out of which houses could be freely generated. This is true even of Gropius' most independent designs, at the Berlin Growing House exhibition.

Although these two examples were acclaimed at the exhibition, Gropius' hope that he had at last produced a prefabricated house "to the point of salability," as he put it, was not to come to fruition.[75] The dream of the factory-made house, apparently within his grasp, was still to prove elusive. Professional acclaim was not enough: what mattered was the failure to achieve commercial success. We must examine the reasons for this disappointing culmination to the copper house venture. An indication that all was not well may be seen in the somewhat acerbic tone of Hirsch Kupfer's letters to Gropius, at the time of the exhibition. There is constant bickering (albeit in a restrained and dignified style) about the erection and exhibition accounts. This pettiness must be regarded as a

4.17
Walter Gropius, Growing House using Hirsch system, showing revised construction details, Berlin exhibition, 1932

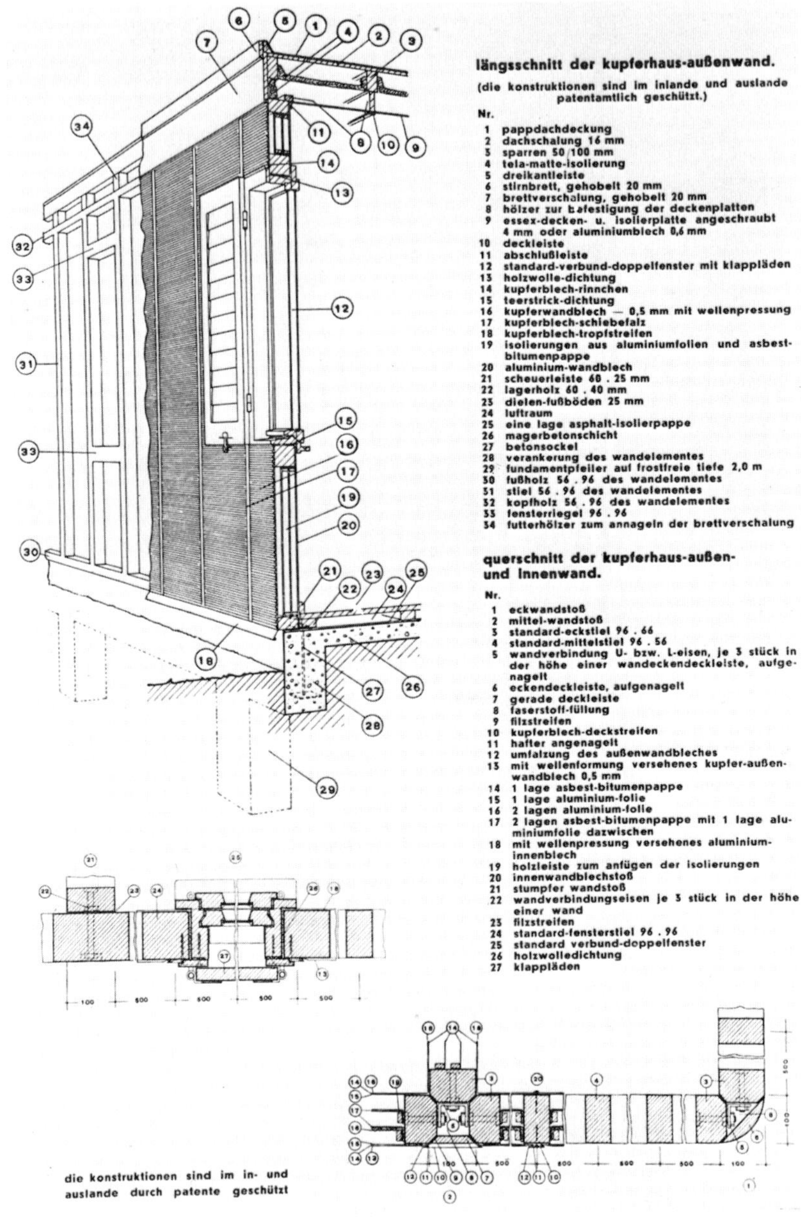

4.18
Walter Gropius, Growing House, using Hirsch system, view, Berlin exhibition, 1932

symptom of stress rather than a cause, for although Gropius often acted independently and on his own judgment, there is no suggestion of extravagance, and his final account for all exhibition works—from bricklaying to sign writing—still came within the RM 8,000 budget allocated.[76] The relationship of Gropius to Hirsch became uneasy, and until Gropius made the express wish, Hirsch Kupfer's directors did not even visit the exhibition stand. On 25 May 1932 a circular notice informed Gropius that a Herr Ebeling had been appointed head of the Copper House Department at Hirsch Kupfer and was to be regarded punctiliously by all concerned as the sole source of authority in all actions concerning the copper houses. The honeymoon with Gropius was over; however, a new romance at Hirsch Kupfer did not appear to be threatening. It soon became clear that Ebeling's function in the Copper House Department was not to develop it further, capitalizing on the triumph of Gropius' tour de force at the exhibition but—surprisingly—to wind it down, as painlessly as possible.

We say surprisingly, and yet with the benefit of hindsight, that the collapse of the copper house venture appears inevitable. Its failure is not the function of personal issues between Gropius and Hirsch, nor technical or aesthetic problems with the copper house. Its proximate cause is financial, and its inevitability stems from the macrocauses of German politics and international economics. In the second half of the 1920s the German economy had made an incredible recovery. Runaway inflation was halted, the currency stabilized, and real economic growth took place. In 1929, when the copper house project was about to be initiated, industry for the first time since World War I appeared soundly based; the Weimar government had achieved a somewhat precarious stability, and prospects for the future looked bright. The Great Depression in the United States, however, soon undermined this optimistic picture. The next three years, 1930 to 1933, are the critical years of our story: they have been characterized as years of "crisis and deflation."[77] Production fell rapidly, and unemployment grew. There were over 5 million unemployed in Germany in 1931, and in 1932—when the tremors of the economic earthquake hit Hirsch Kupfer—over 8 million were jobless. Housing in Germany reached a peak in 1931 (an optimistic year for Gropius and Hirsch), but this represented investments made at least two years earlier, when "the capital gains of industry were increasingly invested in housing." By 1932 the bubble had burst. "House building was cut back although the need for housing among the poorer classes had increased. . . . Demand and purchasing power declined. The number of empty flats increased."[78]

Much more than the housing market was threatened by 1932. The economy as a whole was in jeopardy, and the metal industry—of which Hirsch Kupfer-und Messingwerke was one of the giants—became increasingly vulnerable, and not only to economic threats. The economic chaos was matched by political turmoil. The frenetic struggle for power culminated in the inexorable march of Hitler toward the Chancellery and eventual dictatorial authority, black events which—as we talk of the fate of the copper houses—lie in the months immediately to come. One must remember here that Hirsch Kupfer was a Jewish firm, by which is meant that not only was it controlled and directed predominantly by Jews but that the Hirsch family, the dynasty which shaped it, proudly proclaimed its Jewish identity. The Hirsches of Halberstadt and Eberswalde maintained a prominent Jewish profile, rather than seeking (as had done the Rathenaus of AEG) to assimilate with the mainstream of German society. The Copper House Department moreover was, for Hirsch Kupfer, but one division in an industrial empire with worldwide ramifications, all of which was now threatened. When we see it in these proportions, in the light of the massive economic turndown and with National Socialism now standing on the very threshold of power, the whole copper house venture has for Hirsch Kupfer dimensions of significance necessarily less than those Gropius ascribed to it. Hirsch was beset by troubles on a wide front.

Even as the copper houses were being triumphantly erected in the Berlin exhibition grounds, in April 1932, an examination of the state of affairs at Hirsch Kupfer was being undertaken at the request of the major foreign shareholders, including Haut Katanga and the British Chemitrust.[79] This urgent review revealed a sound technical basis to the firm, as might be expected, but also a shortage of capital of about RM 10 million, which almost equaled the existing share capital. Major reconstruction of Hirsch's financial structure would obviously have to take place immediately. In such a situation the Copper House Department, because of high costs and a shrinking market, could obviously show no short-term prospects of profitability. And it was the immediate future that counted. The financing of copper house production, so optimistically viewed but a year before, had now become not only an embarrassment but an outright liability. Hirsch did not conceal this view from Gropius but wrote to him, on 3 May 1932, telling him with almost brutal frankness that for them the only solution of the problem was one "according to which we shall not have to carry out the manufacturing of the copper houses and their financing."[80]

The decision was taken in principle in May. It became public knowledge a month or two later, when *Bauwelt* magazine published an article, regretful in tone, entitled "Nicht Mehr Kupferhäuser." The Hirsch company, it pointed out, had suffered many financial setbacks during the Great Depression. They were forced to write off many losses, and—as one of many economy measures—they decided at the beginning of 1932 (at the very time the Gropius houses were being assembled) to stop production of the copper houses which were, alas, not profitable. As they explained, "the possibility of selling these houses . . . were very few indeed. The copper houses are not much cheaper than houses produced in the ordinary way, and therefore they have not created a great flurry of demand." The two copper houses developed by Gropius, and so prominently featured in the Berlin exhibition, were thus in fact the last of the run. The hope was expressed, however, that they might "bridge across to a time when there will be a demand for such houses that can be built in workshops, which are easy to transport and which can rapidly be assembled on the building site."[81] This hope was soon to be realized, modestly, and in a strange and oblique manner.

The Search for a Viable Alternative

When the representative of the overseas investors, Dr. Lauber, reported on the parlous state of Hirsch's finances and recommended a basic reorganization, Gropius hurriedly sought from him an undertaking that no new arrangement would be entered into during the next three months. This moratorium was needed, Gropius explained, to give him a chance to complete negotiations he had begun on Hirsch's behalf which were in a sensitive state. Even before the crisis in the copper company's affairs broke, Gropius had been seeking ways and means of putting the financial organization of the Copper House Department on a more solid basis. From the very outset of his relationship with Hirsch he had been advocating the establishment of a separate marketing and finance company, and Hirsch had in principle accepted that view, on the basis of similar successful American models. In January 1932 Gropius and René Schwartz of Hirsch Kupfer had conducted talks with a Herr Moufang, in which they raised the possibility of creating a financial company, with all those interested in the copper houses participating. A capital sum of RM 10 million was mentioned, a large sum for those days. Also in January Gropius began an extensive series of exchanges with Otto Mandl, a Berlin man of business, with whom he had first established contact, on behalf of the Copper House Department, the previous July. In asking for the opportunity to continue these negotiations, he informed Lauber in confidence that "Mandl has made serious promises that he will do everything to bring about the financing."[82] This was at the end of April 1932; Hirsch agreed to give him until 26 June, a grace period of two months, on the condition that they could in the meantime forward any offers of people interested and that Gropius would take it upon himself to investigate their seriousness. Gropius turned to Mandl with some urgency: "I now ask you to look for a suitable way to put the construction of the copper houses on a healthy financial basis." Gropius himself would spare no efforts, he assured Mandl, to help him achieve this end, whether by means of more documentation, travel, or participation in further discussions.[83]

The search for a new parent company now moved to France, and Mandl put Hirsch and Gropius in touch with the French Aluminium Trust, part of the Péchiney Group of Lyons, one of the most important concerns of the French chemical industry.[84] Hirsch Kupfer were asking a considerable sum[85] for the rights to the production method, and a series of complicated exchanges between Paris and Berlin took place to examine the exact situation of the patent rights. The legal department of Hirsch Kupfer spelled out the details of the situation to Gropius: "We inform you that we started the cancellation of the contracts with the inventor of the method for the production of houses of copper sheet with effect from 31 December 1932. We have also canceled the contract with Dyckerhoff for insulation foil, also from 31 December 1932. . . . For the new people who will be interested in the field of copper houses, only the patent of Förster and Krafft seems to be important.'' The sad story of the winding up of the Copper House Department seems to have begun. ''If Mr. d'Auvigny [of the French Aluminium Trust] is a serious interested party, we shall be ready to give him the necessary documentation through our patent department, and we shall inform him of our experience in the field of copper houses. We shall point out that we shall not carry out the production of copper houses any more. However, we shall be ready to put our experience at his disposal, and perhaps we will be ready to rent out to him the existing department for the production of copper houses.''[86] For Hirsch, the decision was unequivocal—they would not be directly involved with the copper houses in any fashion. For them it was the end of the road. And as the arrangements with the French Aluminium Trust failed to materialize, so it proved for Gropius' involvement in this project as well. Despite continued discussions on the obdurate question of patents—Gropius is inquiring of Mandl in September 1932 whether there is any progress with Förster, and we even see him referring to a patent lawyer, on behalf of the German Copper Institute, as late as February 1934—no solution involving Gropius is to be found for the copper houses.

The last act of the drama is played out. By the end of 1932, beset by financial troubles and in an era of growing political tension, the giant Hirsch combine went into liquidation. From December 1932 onward Gropius' correspondence from Finow came not from Hirsch, but the Berlin Ilsenberg Metallwerke A.G., a successor, or holding company perhaps, to Hirsch.[87] Old Aron Hirsch, the patriarch of the firm, was removed from the directorate of the copper concern and involuntarily retired by the Nazis, early in 1933.[88] It was in fact at about this time that all Jewish members of the Executive of the German Metal Industry were removed from office.[89] Only Emil Hirsch's Erze und Metalle Hirsch A.G. of Berlin (and Amsterdam) continued to operate, until the Berlin branch was eventually closed down (or taken over) by the Nazis. The copper industry was of great strategic importance to a Germany illicitly rearming; moreover, as we have seen, the Hirsch family had played a proud and prominent role in German Jewish affairs and were a natural and conspicuous target for Nazi discrimination.

As a footnote to the history of Hirsch Kupfer-und Messingwerke, it is interesting to note that today the company still exists, as a fully owned subsidiary of AEG, the giant German electrical combine.[90] There had long been loose commercial ties between AEG and Hirsch. From about 1922 these links began to consolidate, and by 1927–28 the relationship between the two companies, the one a major producer of copper products, the other a major consumer, had become so close that, in the interests of the rationalizing of production, they shared each other's production facilities. It is not clear from the records, presumably lost in the chaos of war, exactly when—between the reported bankruptcy at the end of 1932 and the re-establishment of Hirsch Kupfer after the war[91]—AEG became the sole proprietors of Hirsch. But there is an historical irony in the situation. It was in a memorandum to Emil Rathenau of AEG (it will be recalled) that Gropius first articulated his visionary concept of the factory-made house, and it was through Hirsch Kupfer that he came close, with the evolution of the copper house, to realizing this dream.

In November 1932, as the family business of Hirsch Kupfer finally ceased to exist, one of its directors, René Schwartz, a leading figure in the Hirsch Copper House Department,[92] took an important decision. Walter Gropius learned of this unofficially and wrote to Schwartz with his characteristic generosity of spirit: "A little while ago I heard that you have taken over the direction on your own of the development of the copper houses. I wish you success and hope that you are on your way to overcoming the difficulties which you had encountered at Hirsch Kupfer-und Messingwerke."[93] The development to which Gropius referred was made public in a circular letter,[94] issued early in December 1932, under the printed heading: "*Deutsche Kupferhaus Gesellschaft*." This letter, addressed to all those interested in the copper houses, announced: "We have the honor to inform you that we have founded a company under the above name, which has taken over the rights and production facilities of Hirsch Kupfer-und Messingwerke for the production of copper houses." An exhibition and sales room was opened at 65 Unter den Linden, and sample elements and models of houses were on display to the public from mid-December. After this initial announcement all references to Hirsch are dropped, but the indebtedness to Hirsch prototypes and practices is everywhere felt.

This is evident in the catalog published by Deutsche Kupferhaus Gesellschaft. The entire construction method, according to the catalog, is based on the Förster and Krafft system, and it is identical to the original (pre-Gropius) Hirsch house. Even the very photograph of the Hirsch house at the 1931 Berlin exhibition continues to be used in the advertising brochure of Deutsche Kupferhaus. The selling effort is directed to Germany itself. The prices are quoted within a delivery radius of 50 km from Berlin, and letters of recommendation from satisfied customers in various German districts are published. Some of these, at least, are from customers of the old Hirsch era. Thirteen house types were illustrated, with prices ranging from RM 16,850 (for the seven-roomed, 164 sq m "Favorit") to a modest RM 3,900 (for the one-roomed, 25.5 sq m

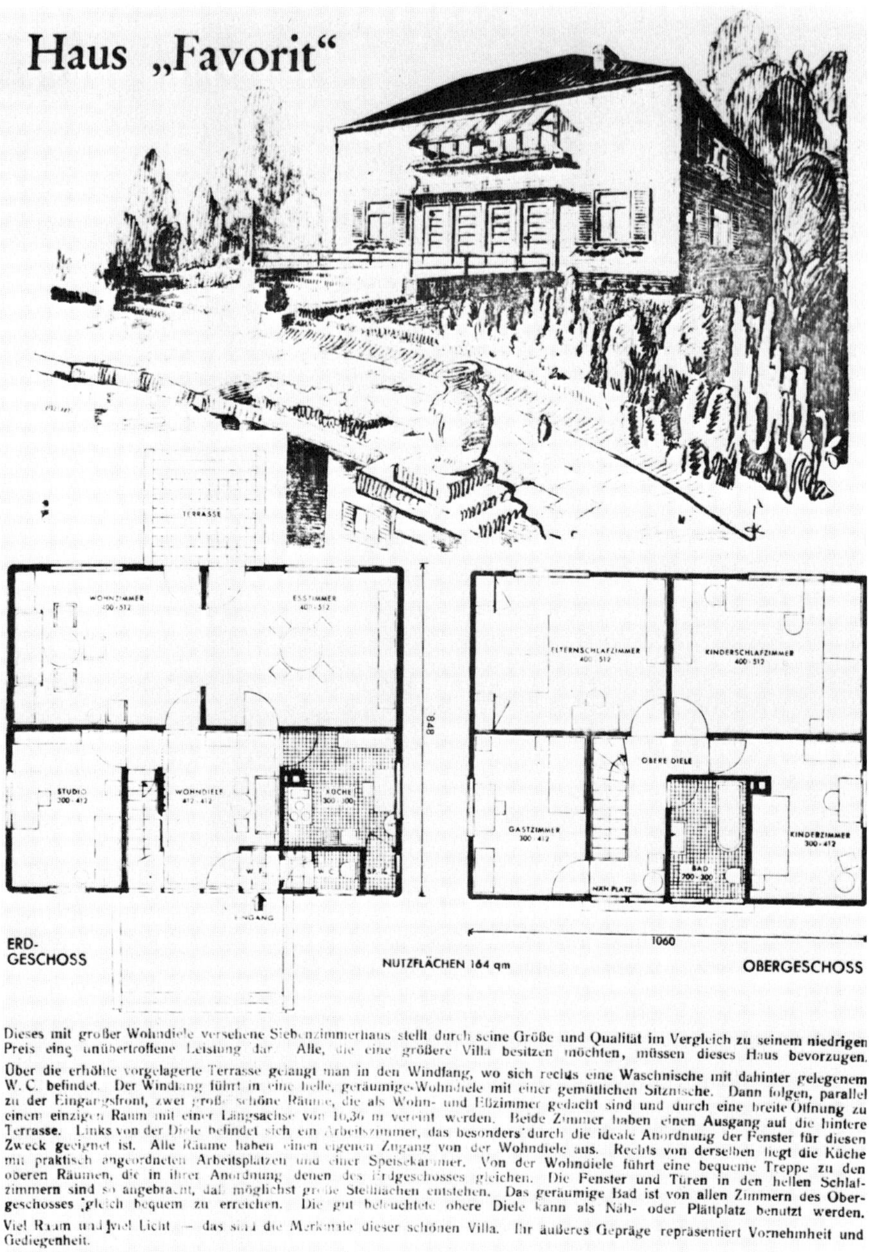

4.19
Deutsche Kupferhaus Gesellschaft
"Favorit" type, 1932

"Neues Leben," a core house designed to expand in stages to an ultimate size of 80 sq m). Of the houses illustrated, many were types originally listed and named in the Hirsch price lists of 1931; others appear to be new and were printed in the catalog with the name Deutsche Kupferhaus Gesellschaft included on the page. We will look in vain, however, for any sign of the Gropius types—the product of so many months of intensive development—or the models that had been featured at the Growing House exhibition. The reason for this surprising omission of the most highly developed of the copper houses was soon conveyed to Gropius, in an oblique manner which only compounded his distress. Either directly, or through Gropius' good offices, several enquiries about the copper houses had come to Schwartz. Some of these referred specifically to the Gropius types. Schwartz's reply to these potential customers was brought to Gropius' attention by one of them. "The Gropius houses mentioned by you," wrote Schwartz, "were only an experiment by Hirsch Kupfer-und Messingwerke, which was, however, not followed up. We produce the old copper houses which found greater acceptance among our circle of clients."[95]

The Deutsche Kupferhaus Gesellschaft had risen, phoenix-like, from the ashes of Hirsch Kupfer. For a while, it appeared as if the saga would continue; but Gropius' connection with the copper house, a product he had labored so mightily to improve, was now finally severed. Gropius viewed the future of this new company with some skepticism, but nevertheless his disappointment must have been deep.[96] However, the fate of the copper house venture was even then being overshadowed by events of a much more tragic dimension. These events not only destroy the Hirsch industrial empire, they engulf the Hirsch family itself. Ironically, they are also the proximate cause of the next, and last, chapter in the history of the copper houses.

Part II

Politics and Technics: Years of Crisis

"The new dwelling is an instrument for the destruction of the family and the race."

"To oppose 'bolshevist' urban architecture, Nazism must . . . encourage young architects to create an architecture based upon 'nature and the soil,' using Handwerk rather than machine technology."

In July of 1931 . . . in a long article on the Reichforschungsgesellschaft, the paper [the Völkischer Beobachter*] devoted several columns of invective to "the members of the Jewish-Bolshevist architectural organization, the Ring" . . . (and) promised that the party would "settle accounts" with these men when it came to power.*

Barbara Miller Lane, Architecture and Politics in Germany, *1968*

By 1930, the American system was in serious trouble. . . . As the economy ground to a halt, more and more people became unemployed; banks were forced to close their doors, and more Americans faced bankruptcy. During the winter of 1930–1931, many homeless Americans simply wandered about the countryside while others began to live in shacks made of scraps of lumber, old boxes, and newspapers. In New York and other cities, public soup kitchens opened, and long lines of hungry people formed before them.

Heritage of Freedom: A Brief History of the United States, *n.d.*

5

The Palestine Prefabs: A Case Study

Germany and Palestine

Our story now turns to deal with the importation into Palestine of a number of copper and other German prefabricated houses of various types, and their erection there, in the early years of the Nazi terror. It is a Janus story with two faces: the one an account of industrialized civilization at its height, of technological skill and ingenuity; the other a tale of cynicism and high drama, encompassing the eviction of an ancient people from Germany at a time when their fellow Jews were struggling to establish a homeland in Palestine, then under British mandate. Technics, economics, and politics are inextricably interwoven in this account of a bizarre episode in the history of prefabrication, an episode nevertheless indicative of the turbulent era of the thirties. Bizarre, yes; indicative of the special circumstances of the thirties, certainly—but also a chapter in the history of prefabrication that reiterates a recurrent theme, the theme of prefabrication as a response to crisis situations. The high points in the long story of prefabrication relate to the unusual, the unforseen, the remote, rather than to the norms of settled communities living stable lives. More often than not, prefabrication has flourished in emergency situations: a new colonial settlement, a military outpost, a mining town, a tornado, a war. More often than not in the past two hundred years prefabrication has also been linked to the concept of export, of the satisfaction by the industrial nations of urgent needs elsewhere. The story of the Palestine prefabs is therefore more than an account of a unique situation; it is a well-documented case study that illuminates a much more general situation and casts light on generic possibilities and problems that characterize the movement toward the factory-made house.

On 30 January 1933, Hitler became Chancellor of Germany and assumed dictatorial powers five weeks later. The position of the Jews in Germany, already threatened by the growing power of National Socialism in 1932, now became more openly imperiled. The possibility of emigration, which some had considered even in 1932, now became increasingly imperative, and many German Jews turned their eyes toward the Jewish homeland. These headlines of the *Palestine Post* of Jerusalem of early 1933 tell the story in staccato form. On 12 February, the almost laconic statement: "Jews Begin Exodus from Germany," is soon followed, on 15 March, by the more poignant "Jews Flee the Nazi Reign of Terror"; and on 16 April we are told that the small, vital, and well-equipped Jewish *Yishuv* in Palestine has begun active "preparations for German immigrants."[1] Jewish settlement in Palestine was strictly controlled by Great Britain, the mandatory power. Immigration was restricted to a fixed monthly quota; however, some additional "capitalist" visas were available for those potential immigrants who could muster the required capital sum of £1,000.[2] It was to these "capitalist" visas that many of the Jews fleeing Germany aspired. But although the more affluent of the well-established German-Jewish community had no difficulty in raising this sum, German foreign currency controls (even prior to Hitler) prohibited the transfer of such funds overseas. To break this impasse, various agreements were initiated in Palestine; and these "transfer agreements'' provide the complex legal and economic framework within which the export of German prefabricated houses to Palestine was to take place.

Negotiations for a transfer agreement commenced privately in 1932, before the Nazis came into power but after the imposition of German currency controls.[3] These negotiations resulted in an agreement being signed, early in 1933, between Hanotea (a privately owned citrus-growing company in Palestine) and the German Ministry of Economics, ''providing for the transfer to Palestine of one million marks belonging to German Jews . . . ''[4] On 18 July 1933, this agreement was extended to three million marks. At this scale it was obviously beyond the scope of a private arrangement, and in August 1933, after extensive negotiations between the Zionist executive, Hanotea, and the German Foreign and Economic Ministries, the Anglo-Palestine Bank (the principal financial institution of the Zionist organization) was recognized by the German Government as ''the competent authority in all matters affecting the transfer of Jewish capital from Germany to Palestine.''[5] Consequent to this agreement ''the Trust and Transfer Office Haavara Ltd. was established in Tel Aviv . . . to facilitate the emigration of Jews to Palestine by allowing the transfer of their capital in the form of German export goods.''[6] In part these transfer agreements solved the problem of the capitalist visa, for ''those who deposited their money for transfer were permitted to purchase foreign currency at the Reichsbank to the value of £1,000, to serve as a deposit with the British immigration authorities.''[7]

For the Nazi government[8] the agreement was important for a variety of reasons, some ideological—such as the desire to facilitate emigration of the Jews and thus make Germany *Judenrein*; some economic—the necessity to encourage exports and thus revive Germany's industry struggling for rehabilitation in the wake of the defeat of 1918, the political and economic chaos of the twenties, and the depression of the early thirties; and some political—the desire to deflect the planned economic boycott, by which liberal forces in the west hoped to moderate Germany's vicious antisemitism. For the Zionists[9] the agreement posed a tragic dilemma—economic cooperation with the hated Nazis in order to save Jewish souls. Despite the crises of conscience and ideological conflicts which these agreements engendered in both the Nazi and Zionist camps, the transfer agreement grew in scope and importance over the years. In the period 1933 to 1939 some 60,000 German Jews emigrated to Palestine, and a total sum of over £8,000,000 (over $40 million) was transferred,[10] the German share of the Palestine market growing from less than 10 percent in 1932 to 16.53 percent in 1937.[11] At first the whole range of German industrial output was included in the agreement, and thus, it is significant to note, building components and building materials. The implications of this for the prospective emigrant to Palestine were great, as was realized by both Germans and Jews. As early as 7 September 1933, the *Palestine Post* pointed out a special provision of the transfer agreement whereby "those who wish to build a home in Palestine may pay sums of up to 50,000 marks per person into another 'special account' which shall also be used to pay for deliveries of German goods to Palestine."[12] Until 1936 (when as a result of its rearmament program Germany limited the exportation of strategic materials, including all metallic goods),[13] this fund enabled a large number of German migrants to Palestine to purchase building materials and components in Germany for use in their new homeland. It also enabled a number to purchase complete prefabricated houses, in component form, for shipment to Haifa, prefabricated houses that were manufactured in Germany and ultimately assembled in Palestine.

It is perhaps appropriate to recall how, two years earlier, the Russians had sought, in their contacts with Gropius, to come to a somewhat similar arrangement. The transfer of prefabricated copper houses to Russia, which they had then proposed, was also intended to solve two simultaneous problems: the housing shortage, and the difficulties with foreign currency exchange. Although the instances are by no means identical, there are intriguing parallels.

The history of prefabrication has been described[14] as a record of successful response to the challenge of recurring crises, when local demand exceeds the local capacity to supply. These enabling conditions certainly applied to the situation in Palestine in 1933. The housing position had already been critical for some years, especially in urban areas, where an occupancy of 2.3 persons per room was further aggravated by the poor, often improvised, condition of the housing stock. "In Palestine even today," an informed commentator wrote in 1933, "a considerable proportion of the urban population is still housed below the minimum standard conditions in temporary wooden barracks, tents, etc."[15] With a rapidly growing urban population, due both to natural increase and a much accelerated immigration flow, expected to reach 35,000 in 1933, it was estimated that at least 17,000 to 18,000 further rooms were urgently required. To meet this demand, a considerable increase in building activity took place—in Tel Aviv and Haifa, building output increased fivefold. The consequences of this building boom[16] were immediate and predictable: building costs shot up; land values soared; a chronic labor shortage was especially felt in the building industry, and skilled tradesmen were at a premium; and, despite a growing emphasis on the local production of building materials, critical shortages generated an enormous increase in the importation of building materials—especially cement, timber, and structural steel—from overseas.[17] Much of this material, as we have noted, was imported from Germany, by means of the transfer agreement. Both in Palestine and in Germany[18] the realization grew that the crisis conditions in Palestine housing produced at least a potential market for prefabricated housing, for which the technical infrastructure already existed, as we have shown, in Germany, and for which the transfer agreement could provide the necessary financial framework. So despite—or because of—the tragic situation that had begun to envelop German Jewry, the strange chapter in the story of the bringing of prefabs from Nazi Germany to the land of Israel began to unfold.

The Copper House in Palestine

The *Judische Rundschau* was a weekly newspaper in the German language directed toward the Jewish community in Berlin. It dealt with matters of local Jewish communal interest and with Jewish affairs on a world scale, with an emphasis on Zionism and events in Palestine. At the end of June 1933, in a column of miscellaneous information, it drew the attention of its readers to the phenomenon of the copper house in Germany, in which, it claimed, there had recently been shown a strong interest. These copper houses had been exhibited at the Berlin Building Exhibition and had also won a Grand Prix at the Paris Colonial Exhibition, thus demonstrating their suitability for subtropical climes. After giving a detailed technical description, the *Judische Rundschau* concluded by stressing that the copper house "is very light to transport, and can be erected in a few days."[19]

We may regard this editorial note as preparing the ground for an advertisement in the same issue—the first of a series extending to the end of the year—inserted by the Deutsche Kupferhaus Gesellschaft.[20] This first advertisement is directed in general terms to the intending emigrant: the copper house will give him a very good capital investment in the fastest and best of building systems, well insulated against heat and cold. In July the advertisements become more specific: take your own copper house to Palestine—they advise the emigrant—and you will dwell in cool spaces despite the great heat. From mid-July therefore the "Kupferhaus" is linked explicitly to the settler in Palestine. We must remember that July 1933 was a critical date in the history of the transfer agreement: the Hanotea agreement endorsed by Germany in March was expanded to the sum of 3 million marks on 18 July, and negotiations were underway for the formal "Haavara" agreement, which was officially endorsed the following month. It is reasonable to assume that the Deutsche Kupferhaus Gesellschaft entered into discussions for the right to export to Palestine prior to the appearance of the advertisements: formal approval, under the currency controls, was granted by the Minister of Commerce on 24 July for the houses to be taken out by emigrants to Palestine, without affecting their right to transfer £1,000.[21] On 10 August, following this permission, a special "Palestine Edition" of the Deutsche Kupferhaus catalog was published, entitled: "Warum Kupferhäuser für Palästina?"[22]

The Palestine catalog contained details of six building types: the Haifa, Jerusalem, Tel Aviv, Jaffa, Scharon, and Libanon models. They do not repeat any of the specific designs in the original catalog but clearly relate to the character of the earlier work. They obviously derive from the same set of components, and possibly from the same, anonymous, designer's hand. The range of styles is narrow, and more consistent, than in the pretentious and eclectic general catalog. The Palestine designs are simple and "contemporary" in expression, eschewing the traditional romanticism and complex forms of some of the original models. The roofs as illustrated—although not necessarily as built—are all flat, with wide overhanging eaves: and only in this respect do we see any echo of the "reworking" of the designs which Gropius undertook for Hirsch, at the Growing House exhibition of the previous year. It is significant that only two of the six models are for one-family houses (the standard house type in the original catalog), and the remainder are multiple dwelling units, ranging from two to four dwellings per building. This probably reflects a policy change resulting from a knowledge of the housing situation in Palestine, with its emphasis on small dwellings for letting purposes attracting high rentals. The ownership of one of these multiple dwellings would then assure the emigrant not only instant shelter but a steady source of income. The needs of the local market, and local conditions, have clearly been taken into consideration and—as the catalog points out—the details and construction were specifically modified to suit Palestine, with respect to solar insulation, fly-screening, and other protective devices. The Kupferhaus company was prepared to make arrangements for shipping and charged a surcharge of 4 percent for packing for sea transportation. Their comprehensive arrangements did not end there, however: the catalog promised that a local agent would be appointed to see that proper specialists were available to erect the houses.

The Kupferhaus company took two important steps to fulfill this undertaking: they appointed an agent in Haifa to take overall responsibility;[23] and they sent out an engineer, Herbert Markowicz, after a special four-month training course at the factory, to be technical supervisor of the erection of the houses.[24] Markowicz arrived in Haifa at the beginning of November 1933, at the same time as the first shipment of houses. By the end of the year several of these prefabricated houses had arrived.

It looked as if the copper house project had got off to a good start, but difficulties were immediately encountered: an unanticipated import duty of 12 1/2 percent was imposed on the houses, causing much concern to the importers, as well as to the Jewish Agency.[25] The Jewish Agency, in both its Berlin and Jerusalem offices, was much involved with the experiment of bringing prefabricated houses to Palestine, although it refused to sponsor any of the ventures.[26] It concerned itself with investigations and analyses of the quality and cost of the copper houses[27] and requested a full on-the-spot assessment by the Organization of German Immigrants in Haifa.[28] These studies brought to light, in addition to the import tax, excessive on-site construction costs due to the steep slopes of building sites on Mt. Carmel; a narrow interpretation by the authorities of the building regulations, largely based on conventional British building bylaws, which led, among other problems, to the copper houses being designated temporary buildings;[29] and conflict over the use of unorganized labor with the Histadrut (the Jewish Labor Federation). Gropius, incidentally, would have appreciated the irony of this situation. In his very first memorandum to Hirsch, he had in his wisdom warned of just such problems, when introducing prefabricated houses into a conservative environment. Two considerations had to be fulfilled, he advised, if the program were not to be obstructed locally: representation on the spot with good connections with the local authorities and involvement of local, organized labor to the greatest degree possible.

The conclusions drawn from their investigations by the Jewish Agency were, on the whole, negative, and their attitude therefore tended to be discouraging. The agency was not prepared, at least at that stage, to recommend the copper houses. This in fact was the policy adopted

TYP VI

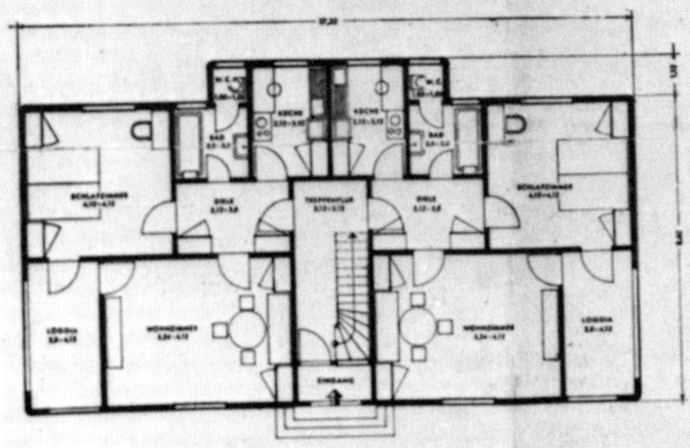

ERDGESCHOSS

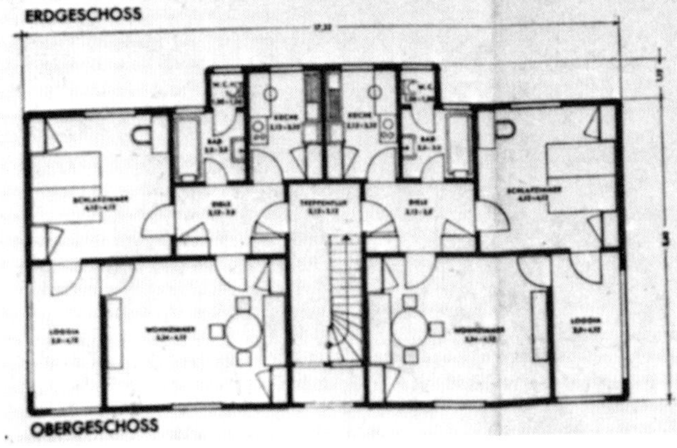

OBERGESCHOSS

Dieser Typ ist ein vier mal Zweizimmerhaus. Jede Wohnung enthält 1 Schlafzimmer, 1 Wohnzimmer mit Loggia, 1 Diele, Küche, Bad getrenntes W. C.

Die Loggia in der 1. Etage kann auf Wunsch offen oder überdeckt geliefert werden.

5.1
Deutsche Kupferhaus Gesellschaft, type VI Dwelling, "Libanon," 1933

5.2
Deutsche Kupferhaus Gesellschaft, Tuchler house, Haifa, construction photo, 1934

by the Jewish Agency to the general question of prefabricated housing from Germany. They did not object to this trade, as they saw no danger in it to the established building industry: on the other hand, with proper official caution, they were not willing to encourage the experiment or endorse any particular system.[30] Perhaps in the light of a total absence of experience of prefabricated houses of this degree of technical sophistication in Palestine—which knew only crude, temporary huts—and the very limited knowledge then available of the performance capabilities of the copper houses, a conservative approach was reasonable. And yet when we consider the long subsequent history of these dwellings in Palestine, their durability and comfort—indeed, the affection in which they were held (and continue to be held) by many of their occupants—then we must hold the caution to have been excessive, the doubts misplaced, and the judgment in error. But this of course is the wisdom of hindsight, nearly half a century after the event.

Estimates of the number of copper houses actually landed at the newly built Haifa Port vary, some accounts putting the total as high as fourteen.[31] Many of these houses were of course multiple dwelling buildings, containing up to four apartments each. The history of nine buildings has so far been traced. Two of these were erected near Tel Aviv, at Ramat Gan (each a large double-storied house) and stood for over thirty years, before being demolished to make way for larger projects, thus bowing to the inexorable pressures of land economics.[32] Special permission, as a concession, was granted for six copper houses in Haifa, of which four still stand; another house exists in Safed. All existing houses are occupied, some by the descendants of the original purchasers.[33] One house—ironically, that of Fritz Neumann himself, the agent of Deutsche Kupferhaus Gesellschaft—was burned down in

Haifa in 1947, and another dismantled in recent years in order to free a valuable building site for more intensive development.[34] Correspondence with the City Engineer's Department of Haifa about the copper houses was entered into as early as December 1933; the first plans were submitted for approval in January; and the first permit to erect a copper house was granted in February 1934. In all, it was a short, concentrated campaign.

The houses are generally similar to, but not identical with, the models illustrated in the Palestine catalog, indicating the degree of flexibility inherent in the system. They have undergone alterations, in various degrees, throughout the many years of their use: additions have been made, internal partitions altered, space under the supporting concrete platforms filled in, in one case a roof replaced. They have now stood for nearly fifty years, continuously occupied, often with a minimum of care and maintenance. Within these limits, and where the original specifications were adhered to,[35] all the existing houses are in remarkably good condition for their age, and they still provide comfort conditions in extremes of heat and cold that make them desirable residences. In the light of this experience, one could today write a much more favorable report than the Jewish Agency was able to do in the atmosphere of uncertainty and doubt of early 1934.

A Diversity of Systems

The copper houses were the first in the field, and they suffered from all the disabilities of being pioneers. They were sufficiently successful, however, to stir other manufacturers of prefabs in Germany to emulate them. With the slowdown of demand at home, many companies were eager to try and exploit the potentialities of the export market, particularly in view of the extraordinary fea-

5.3
Deutsche Kupferhaus Gesellschaft.
Top: Schoenfeldt house, Haifa, 1933–34; *bottom*: Grundmann house, Haifa, 1933–34.

tures of the Palestine situation: its urgency and the special financial arrangements that had been made through the transfer agreement.

In the months immediately following the formalization of the Haavara transfer agreement and the publication of the pioneering Palestine catalog of the Deutsche Kuperhaus Gesellschaft, several advertisements appeared in the *Judische Rundschau* for other systems of prefabrication considered suitable for Palestine. From these and other sources we learn of several examples in metal, some in wood, and at least one proposal, by the engineer Fritz Schlesinger of Charlottenburg, that offered a transportable house for Palestine of reinforced concrete.[36] Timber houses—"transportable[37] *Holzhäuser*"—were advertised by Oscar Lembek (of Leipzig) and Sigmund Korber. Their "Komfortabler Palästinatyp," as illustrated in a sketch, was a substantial-looking double-storied house.[38] We know from other sources that several German timber prefabs came at this time to Palestine.[39] In Haifa there still stands at least one two-storied building erected of timber components (modular panels, faced with horizontal boarding) imported from Germany. This is the Naumburg house,[40] erected in 1938 but dating from an order given in 1935 to the firm of Christoph and Unmack, of Niesky, by Mr. Max Naumburg, a resident of Breslau.[41] By mid-January 1936 Naumburg had received a detailed quotation for the house and its transportation f.o.b. Hamburg. Specifications and drawings for the "Wohnhaus für Palästina: zerieg-und versetzbar System 'Doecker' " were prepared by Christoph and Unmack, based on the design proposals of Naumburg's own architect, in March 1937. The following year, his preparations completed, Naumburg put his emigration in process. In Haifa, the architects Komet and Rath made application to the City Engineer for "permission for the erection of this construction on Mt. Carmel. . . . The house which is due to arrive within 3 weeks time, is a solid construction of wood as they are now popular in Germany and United States, the Northern countries and are exported in great quantities all over the world especially to the tropics. . . . We further enclose a book about modern dwelling houses and public buildings in Germany. . . ."[42] This book, submitted as testimony of the respectability and wide acceptability of advanced timber construction for housebuilding, was *Holzhausbau*, written by Konrad Wachsmann, who had for some time, as we have already recounted, been chief designer to Christoph and Unmack. The wooden building, containing four apartments, was erected according to the instructions and erection diagrams provided by Christoph and Unmack[43] and is still in use today, in part occupied by the original owner himself. The claims of durability made by the applicants, supported by the evidence of Wachsmann's book, have been substantiated by the satisfactory experience of nearly half a century.

In addition to these wooden houses there was significant trade in imported steel-framed buildings. The Germans' interest in steel house fabrication goes back to the 1920s, with the emphasis then, naturally, on the local market. In the 1930s the interest shifted to the export market. In August 1933 the engineer Adolf Locker, in an article entitled "Metallhäuser für Palästina," draws the attention of the *Judische Rundschau* to this long-standing interest in metal houses.[44] He quickly moves from the general to the specific: because metal houses (or at least, as it soon appears, the one particular make he has in mind) are well insulated, they are particularly suited for erection in Palestine. He discusses two types: a "mixed" system, and an all-metal house with factory-made panels. Such a system is versatile, and schools and factories could be constructed, using its components as well as the "transportable *Haus*" for the emigrant to Palestine.

Locker himself would undertake the actual erection of these houses; but for general information, the reader was directed to the architect, Walter Kretschmer, of Berlin. The following month, the Transformbau-Kletzin Co. of Berlin advertised[45] a transportable steel house for Palestine which, it was claimed, was suitable for all situations, and could be quickly assembled with low erection costs. The house was described as being of "modern appearance," an appellation borne out by the accompanying illustration. The Kletzin Steel House was the center of attraction at the Leipzig Spring Fair of 1934, which attracted many visitors. An article in the *Berliner Lokalanzeiger* of 25 March 1934, entitled "German Steel Houses for Tropical Lands," pointed out how necessary it was to revive German exports in order to expand German industry. Houses such as that designed by architect Reichel, and built by Kletzin G.m.b.H. of Berlin, played an important role in this effort.[46] According to the U.S. Department of Commerce, Ludwig Kletzin was greatly interested in exporting his system, based on a 1.22-m modular steel sandwich panel, to the United States.[47] It was of course in the framework of this export drive that the exportation of prefabricated houses to Palestine was pressed by German manufacturers.

The files of the Jewish Agency in Jerusalem expand our knowledge of the German manufacturers of prefabricated houses who aimed at the Palestine market.[48] Georg Breslauer of Berlin, for instance, had approached the Agency's Berlin office in March 1934 with a proposal for a house of steel and aluminium, produced by Grundlisslose Metallhäuser, Leo Szalet, Berlin. The proposal, fully documented, was sent to Jerusalem for the Jewish Agency's consideration.[49] Correspondence continued until August 1934, culminating in a proposal to build four houses in the spring of 1935, which the Agency should endorse if

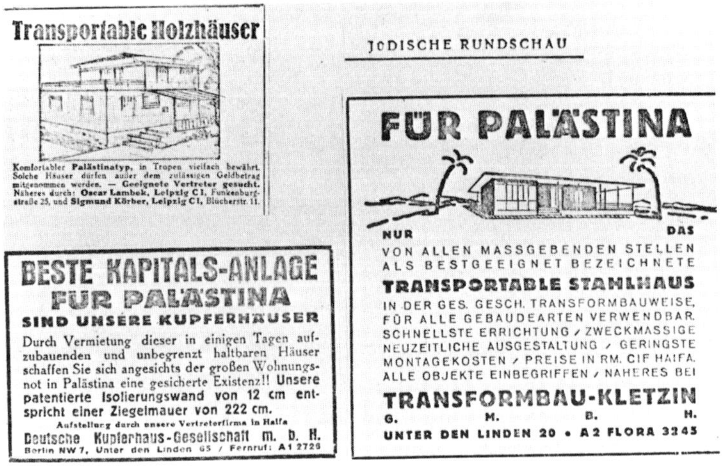

5.4
Advertisements for prefabricated houses for Palestine, 1933

5.5
Christoph and Unmack, Naumburg house, Haifa, 1937–38

5.6
Christoph and Unmack, Naumburg house, Haifa, plans, 1937

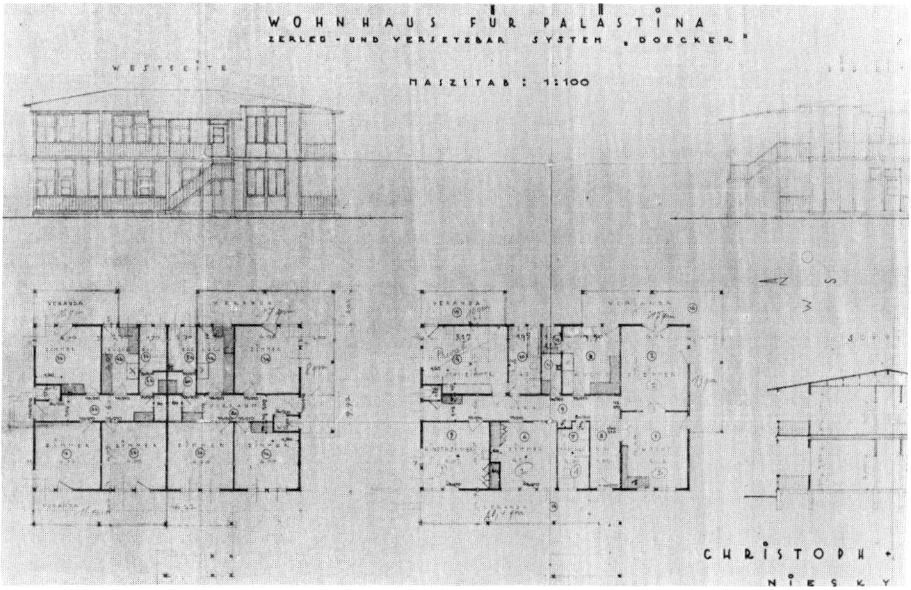

the experiment proved successful. But, as we have seen earlier, the Agency did not feel justified in recommending any specific system.

We do not have any firm evidence that out of all this activity, any steel houses produced by Kretschmer and Locker, the Transformbau-Kletzin Co., or Grundlisslose Metallhäuser, were actually exported to Palestine.[50] The steel prefabs we know to have been erected in Haifa and Jerusalem all come from another and more substantial source: Böhler-Stahlbau, of Berlin. The initiative to export the Böhler system to Palestine was taken either late in 1933, or in 1934, when the Böhler company received authorization from the Economic Ministry to export, under the transfer agreement, to Palestine. A company, the Export Bau-und Handelsgesellschaft M.b.H., was set up in Berlin, which was licensed to handle the exportation of "Böhler Massivbauten" to Palestine. In Tel Aviv, an associated company, the Palestine Building Syndicate Ltd., was formed, to deal with the receiving end of the transaction and to handle the erection of the buildings in Palestine. In December 1934 the Export company published a catalog[51] that gave a technical description of their steel-framed, solid-construction buildings, spelled out the financial complexities of the transfer (60 percent to be paid in marks in Germany, the remainder in Palestine pounds to the Tel Aviv company that could arrange a mortgage), and estimated a conservative eight-month period for design, fabrication, shipping, erection, and completion. It illustrated five standard building types for the Palestine market, ranging from the four-family "Tel Aviv" apartment block (capable of extension to six flats in a three-story construction), to the resplendent two-story "Carmel I" villa. These houses, it was claimed, were designed after months of study of Palestine requirements and conditions; they were flat-roofed structures with projecting hoods over windows and balconies, giving a restrained Mendelsohnian effect.

In Haifa, the German-Jewish architect Bruno Kalitzky—formerly of Chemnitz, but now in partnership with Dov Entin (Eitan)—was familiar with the Böhler system and brought its advantages to the attention of potential customers among his clients. When agreement was reached to use the system, Kalitzky and Entin first designed the project according to the client's needs, then sent the plans to Böhler in Germany, where they were modified slightly to adapt them to the system. Böhler then produced the appropriate set of components, which were shipped to Haifa. Together with the construction kits, they sent out a foreman of works, a Mr. Heimann, to train a local team of artisans in erection procedures and to supervise the construction. Between 1934 and 1937 about six of these buildings were built in Haifa.[52] One, for Heimann himself, was an apartment building with four flats axially but not symmetrically disposed about a central staircase, two to a floor, with an additional small basement apartment designed and built in 1935–37. The other existing Böhler structure of which we have confirmed records is a large block of flats on Hillel Street, in Haifa's Hadar district, probably 1936–38, containing eight flats in all, in a three-story structure. From the outside, seen from the street, these steel-framed imported buildings are indistinguishable from their neighbors (which makes their identification so difficult). Their prismatic forms, simple plastered facades, flat roofs, steel fenestration, projecting balconies with metal railings, the occasional use of porthole windows and vertically glazed staircases are all characteristics typical of the many buildings of the Bauhaus vernacular that were springing up in Haifa and Tel Aviv in the 1930s.

5.7
Böhler Massivbauten, "Tel Aviv" type building, 1934

5.8
Kalitzky and Entin, Levin building, using Böhler system, Haifa, 1937

BÖHLER
Massivbauten
für Palästina

Typ „TEL-AVIV"

Entwurf und Zeichnung sind unser geistiges Eigentum. Ausführung nur mit unserer Zustimmung zulässig.

Nutzfläche einer 3-Zimmer-Wohnung

Wohnzimmer 1	4,80×3,54 m	= 17,0 qm
Wohnzimmer 2	3,50×3,50 m	= 12,3 qm
Schlafzimmer	3,69×3,54 m	= 13,1 qm
Küche	2,81×2,80 m	= 7,9 qm
Bad	3,69×1,90 m	= 7,0 qm
Diele	3,50×1,25 m	= 4,4 qm
Vordere Loggia	4,11×2,20 m	= 9,0 qm
Hintere Loggia	2,68×1,70 m	= 4,6 qm
	Zusammen je Wohnung:	75,3 qm
Gesamtnutzfläche d. Hauses bei 4 Wohnungen:		301,2 qm

Die belaute Fläche dieses Hauses beträgt 179,87 qm, sodaß der Typ Tel-Aviv bei einer zulässigen Bebauung von 40% ein Grundstück von ca. 450 qm = 0,45 Dunam (1 Dunam = 1000 qm) erfordert.

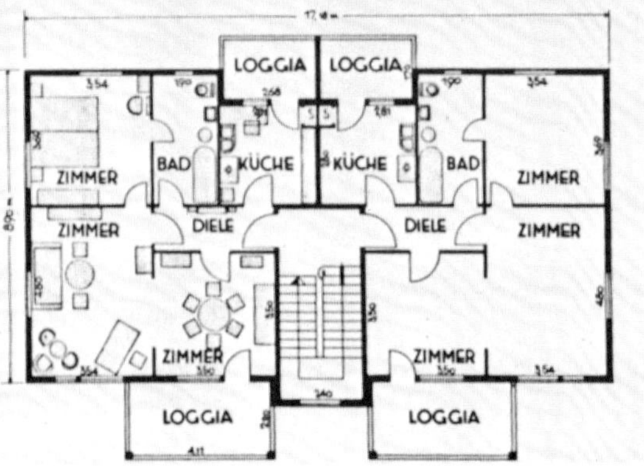

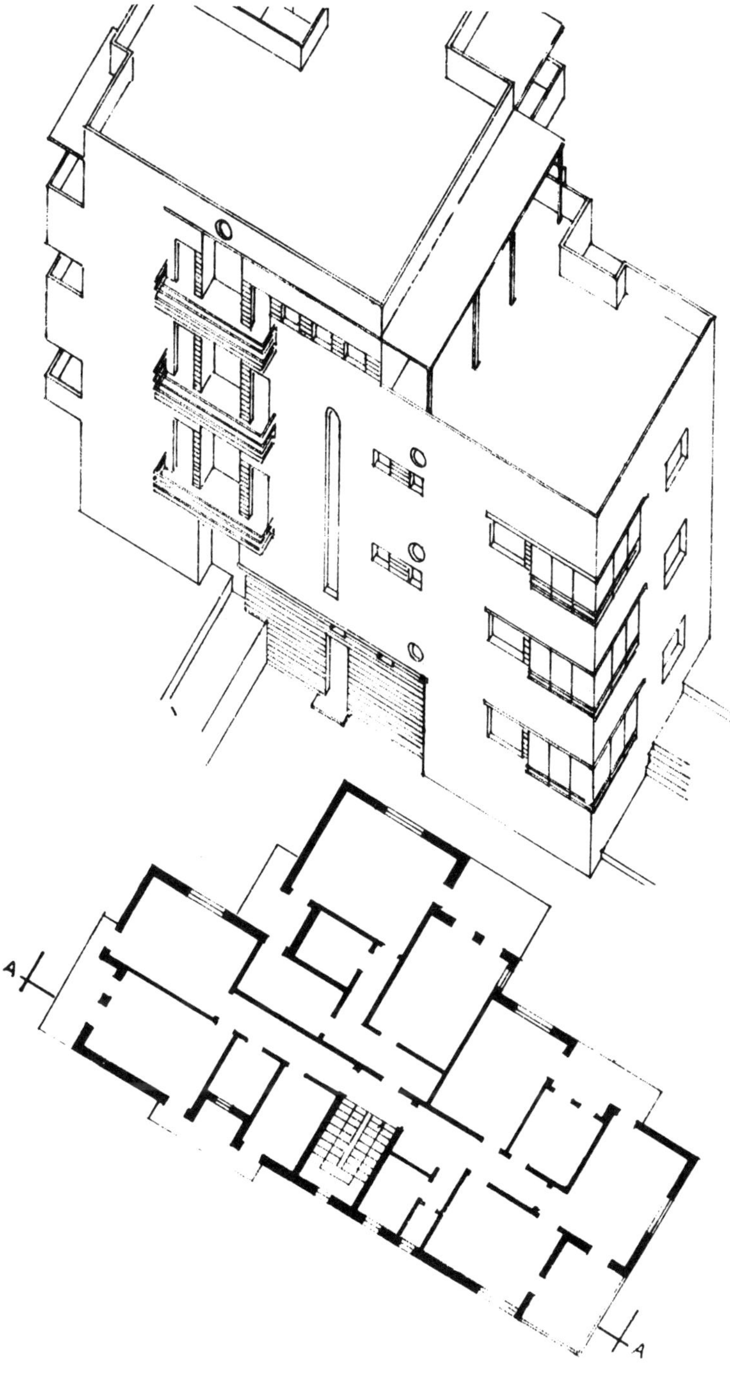
A
A

The modern German influence on the architecture of the Jewish settlement in Palestine at this time was considerable, in both functional planning and austere appearance. But in the examples we are considering, this influence extends into the very structure itself. The system used in Haifa was the *Stahlskelettbauweise*, a lightweight steel frame utilizing for beams and columns U and I profiles, 10 cm in depth. Internally, lightweight porous blocks were laid between the close-spaced columns, with a coverplate concealing the column; externally, the columns were concealed by an outer skin of 5-cm pumice slabs rendered with a plaster finish. Internal partitions were also of these slabs, plastered. The roof construction was of I beams, with a concrete roof over, cast directly onto corrugated iron sheets—a typical Böhler detail; below an expanded metal ceiling, carried on clips, provided the surface for a plastered ceiling. The planned aluminium foil insulation was for some reason omitted. Sanitary installations, normally prefabricated by Böhler, were in Haifa not integral with the system because local building regulations demanded total accessibility. In all, the system as used by Kalitzky and Entin is similar, not to the three Böhler examples of the Growing House task force but rather to the later, more substantial systems illustrated in November 1932 in *Bauwelt*.[53] In addition to these buildings of Kalitzky and Entin, the Böhler option was explored by the Haifa architect Max Loeb, in association with the engineers Weltsch and Heinemann, and a series of sketch plans of a modular nature were prepared of apartment buildings of various types, some variants of the basic types in the Palestine catalog. Loeb possessed a copy of this catalog, and there were contacts with the Palestine Building Syndicate Ltd., with the client, Frau Levy-Stern, and with her family in Germany.[54] This project, as far as can be established, failed to materialize.

In Jerusalem we know of at least three buildings (designed by the architect Dov Kuzinsky) that utilized the Böhler system.[55] Indicative of the sophistication and complexity of these Böhler-based houses is the large building of Dr. Walter Katz, erected in Rehavia in 1937.[56] Above a solid basement (later converted to a flat) there are two typical floors, each comprising three apartments, the building being capped with a roof apartment; that is, the building, originally containing seven apartments, was in part four stories in height. Kuzinsky was apparently familiar with the Böhler system; he was moreover in touch with the Palestine Building Syndicate and its Berlin associate, Export Bau. With the client's agreement, it was decided to handle the importation of the building components directly through these firms, and not through the general arrangements of the transfer agreement, which (because of the falling value of the mark) was by now subject to a considerable surcharge.[57]

The Jerusalem architect prepared his plans, with some rudimentary indication of the steel framing; further drawings were then made in Germany, and all the components were prepared to suit the specific needs of the plan. These building elements, which included the patented Böhler steel skeleton system, the steel-framed stairs, the pumice blocks and slabs for external (double) and internal (single) walls, the steel door frames and roller shutters, the electrical, plumbing, and sanitary installations, even the oil-fired boiler and the central heating system (still functioning effectively today), were shipped out in large packing cases to Haifa and then transported overland, 160 km or more to Jerusalem. There the construction took place simply and speedily—"like a Meccano set," according to Dr. Katz. A German engineer sent out by the Berlin company supervised the erection of the house for six months, then handed over to Mr. Karl Rosenthal (local manager of the Palestine Building Syndicate) who was re-

5.9
Kalitzky and Entin, Levin building, using Böhler system, Haifa, construction photo, 1937

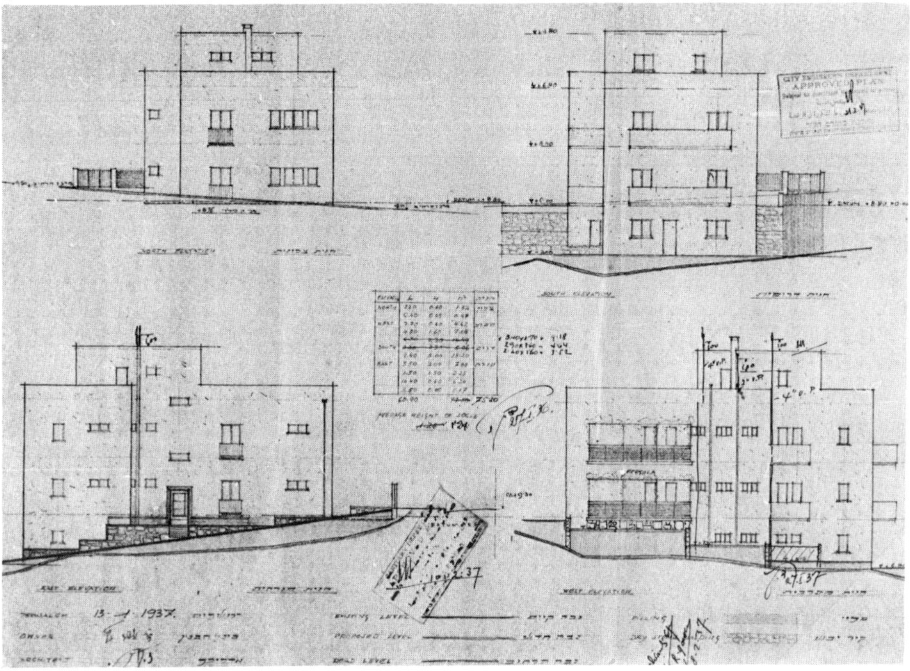

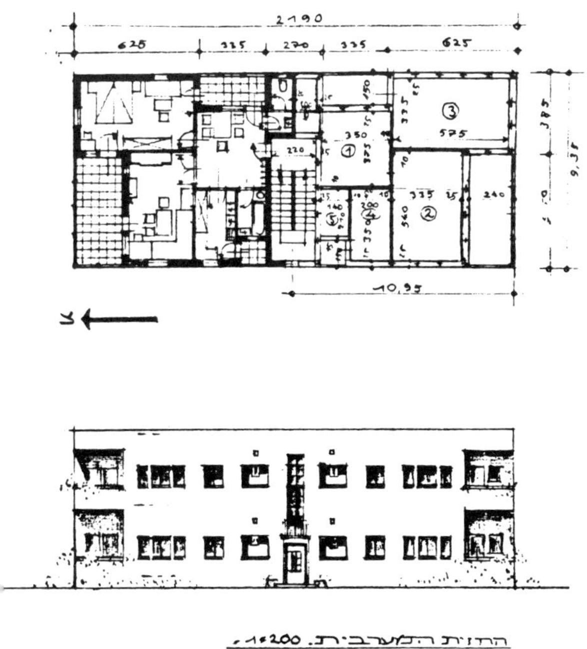

5.10
Max Loeb, with Weltsch and Heinemann, apartment block using Böhler system, c. 1935

5.11
Dov Kuzinsky, Katz building, using Böhler system, Jerusalem, elevation, 1937

sponsible subsequently for the supervision of all three buildings in Jerusalem. Finally, in addition to these examples in Jerusalem and Haifa, we know of two three-story blocks of flats in Holon, built according to the Böhler system in 1935.[58]

As far as can be established, the importation of copper houses ceased in 1934, steel-framed buildings in 1937, wooden houses early in 1938. There can be no doubt that this trade was impeded by real (or perceived) difficulties, in both Palestine and Germany. In addition to doubts about the durability and climatic performance of the copper houses, for instance—doubts which, as we have seen, history has shown to be unfounded—there were many obstacles to be overcome by the pioneer importers. Among these were the unexpected import tax, the problems caused by local building regulations, the misunderstandings with the labor unions (a perennial problem where industrialized buildings are implanted in the framework of a traditional, and conservative, industry), and the disappointment and inconvenience of being granted only a temporary building license. Perhaps most serious of all, the unexpectedly high cost of erection, particularly on the difficult steeply sloping hillside sites of Mt. Carmel, undoubtedly acted as a serious deterrent.[59] Whether these negative factors caused a substantial drop-off in orders we do not know, but we do know that the Deutsche Kupferhaus Gesellschaft was experiencing difficulties in 1934.[60] Some time after 1936 the export of all metallic goods from Germany under the transfer agreement was prohibited,[61] and a selective restriction was probably in force long before then. It seems highly likely, for instance, that the German government, perhaps as early as 1934, would have acted to seal off the outflow of copper, a material of high strategic importance to the munitions and communications industry; thus the government would act to impede the exportation of copper houses. At this time severe restrictions had been placed by the Nazi government on the internal deployment of strategic building materials, with an immediate effect on the prefabrication industry. For whichever of these reasons, or combination of reasons, it seems that the trade in copper prefabs ceased early in 1934, and about three years thereafter the exportation of steel buildings also came to a halt. In any event, as the shadows of 1938 darkened, the time was approaching for the complete cessation of commercial intercourse between Nazi Germany, and the democracies, including Britain and its mandated territory, Palestine.

From a purely quantitative point of view, that is, considered drily as a commercial transaction, the exportation of German prefabricated buildings to Palestine in the 1930s must be considered a minor affair. All in all perhaps only 20 or 25 buildings containing about 100 dwellings found their way from the Berlin factories to their sites on the hillsides of the Holy Land. This could by no means be considered an important factor in the rehabilitation of German industry, nor did it contribute in any meaningful sense to the solution of the acute housing problem in Palestine. Yet, as a chapter in history, it is not without significance. It is inherently interesting because it represents one of the few documented case histories of prefabrication from the user's point of view. Here we see not only the motivations of the designers and producers but also something of the initial hopes, the trials and tribulations, and eventually the experience and fulfillment of actual use for those who seek to solve their own emergency problems through the purchase of a factory-made house. Most of our information about prefabricated houses comes from the architects' magazines or from the producers' advertising brochures. Here the picture is rounded off with a rare view from the other side, with what is almost a consumer's report. Gropius would have learned much about the copper houses, had this information been available to him at the time. He was sufficiently open-minded to learn from criticism; and he would certainly have savored the intricate meshing of events. What a rich tapestry of human interest is revealed in this episode, what diverse threads are paradoxically brought together in its weaving: the dark strands of vile oppression and the shining gold of great technical achievement, black despair and bright white hope, all on a gray field of political compromise and economic opportunism. It serves to illustrate with great clarity the multifaceted and multivalent nature of any event in architectural history.

Prefabrication here becomes more than a philosophy, more than a mere technical episode in the evolution of architecture. Swept up by world events, it becomes—for a small number of people—an instrument for human well-being, perhaps even an instrument of salvation. The chief actors in our drama of prefabrication, Walter Gropius and Konrad Wachsmann, were unaware that the industrialized houses with which they had been associated—the Hirsch copper house and the Christoph and Unmack wooden prefab—had played a role in the poignant struggle of the German refugees to make a foothold in a new land. Despite the long and turbulent history of prefabrication as a crisis response, they had cast for the factory-built house a more fundamental housing function in a more orderly and stable world. Now that world was collapsing, and their own destinies were swept up in the whirlpool of events.

6

Into Exile: With Gropius and Wachsmann to the New World

National Socialism and the Exodus of Talent

Upon the assumption of power by the National Socialists in Germany in 1933, there followed an event unprecedented in the history of modern civilization, an event whose ultimate dimensions of personal tragedy have tended to overshadow and obscure its enormous cultural import in all fields of creative human endeavor, and not least in architecture. We are talking of the exodus, voluntary or enforced, of the cream of German intellectual and creative talent, its widespread dispersion throughout the world, and the cultural dislocation and the cultural cross-fertilization inherent in this dispersion, in what Sibyl Moholy-Nagy wryly called "the Diaspora"—a term normally applied to the exile of the Jews, but which she here chose to give the wider connotation of the "scattering of the faithful."[1]

The cruel weight of Nazi policy and practice fell first, and most heavily, upon the Jews. By 1933 the first measures had been taken "to foster Jewish emigration from Germany and Austria," which by 1941 were to result in more than half a million leaving.[2] Some of these "measures" were cooperative, if not disinterested, as we have seen in the case of the Palestine prefabs; others were of a more coercive and increasingly brutal nature. As a result of these oppressive measures, wrote historian Cecil Roth, "within little more than a year upward of seventy thousand German Jews had left the country. The majority were not artisans or merchants, as might normally have been expected, but professional men—University professors, physicians, surgeons, lawyers, art experts, writers, journalists: in many instances men of international reputation, who had given up their best years and devotion to Germany's service. By the end of 1933 there can have been few countries in the world where some famous German-Jewish scientist or scholar was not at work."[3]

But it was not only the Jews who were leaving Germany. Their massive exodus was accompanied by a sporadic flow of emigration, of smaller proportions, it is true, but of tremendous cultural significance, of German gentiles who found Nazi Germany intolerable. These were the artists and intellectuals, usually of liberal or socialist political persuasion or association, who for motives of conscience, professional frustration, or the sheer necessity of survival in the face of such a harsh regime, chose, or were forced, to leave their native land and culture and expose themselves to the "agonies of the creative mind made homeless."[4]

Prominent among both groups of exiles—Jewish and non-Jewish—were the architects, and especially those architects who had constituted the vanguard of the Modern Movement. There were many reasons for this. As we have already seen, in the closing years of the Weimar Republic unemployment among architects was endemic as the effects of worldwide depression were increasingly felt in the German building industry. Opportunities for employment were manipulated, and in selected instances further restricted, when the Nazis came to power. Jewish architects were excluded from professional practice by the discriminatory membership requirements of the regulatory body, the Reichskammer der bildenen Künste; radical architects of non-Jewish origin, on the other hand, although not "legally deprived of the right to practice . . . nevertheless received no new commissions after 1933."[5]

Among the Nazi ideologists, although there was far from a consensus on this issue, there was a growing tendency to join with such architectural reactionaries as Schultz-Naumburg and his circle, in their attack on the new architecture. The Modern Movement, as a facet of the International Style, had generated a considerable amount of hostility in chauvinistic, conservative circles; this view was expressed not only in the party press but in professional journals. Many of the new architecture's most prominent achievements moreover were in the arena of public housing, cooperative ventures in socially inclined municipalities, and in attacking the architecture, the Nazi propagandists were attempting to denigrate, if not demolish, these potent symbols of socialist achievement.

But this hostility was directed even more to the architects than to the architecture. Most of the prominent architects of the avant-garde were, in Nazi eyes, irredeemably tarred with the socialist (hence, "Bolshevist") brush. The Modern Movement's left-wing origins, in the *Novembergruppe* and the *Arbeitsrat für Kunst*, had not been forgotten, nor forgiven, by the National Socialists. Der Ring, that organization formed by Gropius, May, Taut and Wagner in 1926 to protect the position of the modernists within the architectural profession, was looked on with deep suspicion, especially due to its international links through CIAM, as an organization subversive in the political as well as the artistic sense. And then there was the Russian connection.[6] Many of the leading German architects had always expressed an open interest in, if not a sympathy for, the experimental new society in the Soviet Union. In the 1920s and continuing into the early 30s, the radical

German architects—Gropius, Mendelsohn, Meyer, Poelzig, Breuer—had actively participated in Russian competitions and projects: the Leningrad factory of Mendelsohn, the Palace of the Soviets competition, the Kharkov State Theater. Then, probably most damning of all, from the Nazi point of view, was the actual presence in the Soviet Union of Ernst May, Hannes Meyer, and Bruno Taut, with their teams of German architects, directly participating with their Russian colleagues in the design and planning of the cities of the socialist "new world." Of these architects very few in fact were Jewish, but it helped in their vilification if the racial label (hateful in Nazi eyes) were falsely attached to all of them (if May, for instance, were caricatured as a Bolshevist, complete with Jewish skullcap), so that a hysterical propaganda campaign would be waged against the so-called "Jewish-Bolshevist" nature of radical architecture and its principal proponents.

Under this intense pressure, economic, political, racist and architecturally reactionary, the exodus of the German architects began.[7] Erich Mendelsohn, sensitive to peril and doubly vulnerable as a Jew, was one of the first to leave Hitler's Germany, and in 1933 moved to London. There he was joined, within the next few years, by Walter Gropius, and his former Bauhaus colleagues Marcel Breuer and Moholy-Nagy; by the architectural historians Nikolaus Pevsner and Rudolf Wittkower; by Arthur Korn and several other architects, in a small but highly esteemed coterie of architectural émigrés.[8] London, for many of these, was a congenial but temporary way station before threatening war clouds and beckoning opportunities drew them on to the United States. There, Bauhaus design teachers Josef Albers and Herbert Bayer, and photographer Walter Peterhans, had already settled, and by 1938 they were to be joined not only by the Bauhaus people from London but by Mies van der Rohe and

Ludwig Hilberseimer, who, together with Martin Wagner, had left Berlin to emigrate to America.

Erich Mendelsohn, who had personal ties and professional connections with the land of Israel (including the Schocken family for whom he had built a notable chain of department stores in Germany) commuted between England and Palestine before settling down for a few highly creative years in Jerusalem. In Palestine there had by this time gathered an extraordinary concentration of architectural talent, the only group of exiles from Germany who were at once no longer in exile but at home, emotionally speaking, if not culturally. Alexander Klein, pioneer analyst of functional dwellings and a major collaborator in the development of the *Siedlung*, came to Haifa where he taught for many years at Israel's Institute of Technology, the Technion. Also to Haifa came Adolf Rading, a non-Jew, after he and Hans Scharoun had been dismissed from their posts at the Breslau Academy. Bauhaus graduates Arieh Sharon, Shmuel Mestechkin, and Munyo Weinraub returned home to join a large number of recently emigrated German-Jewish architects of advanced ideas, such as Joseph Klarwein and Heinz Rau. Little wonder that, even till today, Haifa and Tel Aviv have areas that are almost museum precincts of "Bauhaus vernacular."

Of the German architects in Russia, all were disillusioned by the deteriorating situation as Stalin ever more firmly seized the reins of power. Meyer returned to Switzerland. May and Bruno Taut were refused reentry permits into Germany but left Russia nevertheless, May eventually to spend many years in East Africa, Taut to live, until his death in 1938, in Istanbul.

This outflow of architectural talent, which we have briefly sketched out here, had two immediate, connected, and

6.1
Bauhaus faculty, Dessau, 1926. *From left to right*: Albers*, Scheper, Muche, Moholy-Nagy*, Bayer*, Schmidt, Gropius*, Breuer*, Kandinsky*, Klee*, Feininger*, Stölzl* (in 1931), Schlemmer. Those who left Germany, 1933–39 are marked by an asterisk.

6.2
Tel Aviv, general view, showing influence of Bauhaus architecture, 1933

predictable results. The first was a dramatic dimunition of German creative power,[9] for though occasional works of architectural merit continued to be produced, even under the National Socialists, the heart and spirit went out of the Modern Movement at whose center German architects had once stood. Conversely, however, there was a massive infusion of the radical spirit into the architectural Diaspora, which transformed architectural theory, teaching, and practice in countries as far apart culturally as the United States, on the one hand, and Palestine, on the other.[10]

In this period of enforced cultural transfer there occurred a shift of emphasis in technological innovation from Europe to the New World. Germany after a decade or more of intense activity loses its central creative role, and in many fields, including that of the prefabrication of the dwelling, the center of gravity passes to the United States where in the late 1930s and early 40s many interesting new developments in prefabrication are taking place, within a climate of support engendered by the swing from depression to economic recovery. It is to these developments that we must shortly turn. Before so doing, however, we must return to Germany and our two main protagonists, Walter Gropius and Konrad Wachsmann. We must trace their troubled paths as they move from their native land into exile, first in Europe, eventually and almost inevitably, in the United States. In 1941, on the eve of America's entry into the war, they are to come together in creative partnership in New England. The paths that bring them to Lincoln, Massachusetts, are entirely different, and each is, in a way, characteristic of the man. Here are two men, from the same country, each vitally interested in the prefabrication of the house, who do not know each other. Each takes his own course, works out his own destiny. Only once prior to 1941 do those paths intersect in time and place, accidentally and significantly.

Into Exile: The Years of Wandering

As we have seen, by 1932 the effects of the depression were being seriously felt in Germany, especially in the building industry and the architectural profession. Wachsmann had chosen an inopportune time to leave the security of his job at Christoph and Unmack, and despite his *success d'estime* in being the architect of Einstein's house, he managed to secure very little work in his private practice. It is true that after the publication of his book *Holzhausbau*, Wachsmann's name began to be more widely known, especially in architectural circles. In 1930 his studies in timber were reviewed in *Wasmuth's Monatshefte,*[11] in 1931 in *Moderne Bauformen,*[12] and at the end of 1932 he received considerable publicity when *Bauwelt* magazine published his work in a major article on new techniques of building in timber.[13] But fame within the profession does not always signify business success, even in the best of times, and while his reputation increased, Wachsmann's personal position grew ever more desperate. The darkening political situation and the increasing strength of the National Socialists aggravated the exposure and vulnerability of his position, as a Jew. Then, a way out of his dilemma appeared: at the suggestion of his mentor and patron, Poelzig, he applied for, and won, the coveted Prix de Rome. Late in 1932 he left Germany to take up residence in the German Academy in Rome. This, however, was to be but a temporary haven: as the Nazis eventually took over in full power at home, his position in this official institution soon became untenable. Forced by these circumstances to give up his studies at the Academy, he decided not to return to Germany,[14] and his years of wandering began.[15]

In 1933 he traveled in Spain, and for a while acted as assistant to the City Planner of Granada. He then returned to Rome, in 1934, to make his base there for the next few years; but he traveled incessantly, as much as his straitened financial position would permit. His journeys took him all over Europe: to France and Switzerland, Yugoslavia and Greece, Belgium and Holland. He earned little, spent what he had, and lived life to the full. In part because he needed the money, in part because he enjoyed the task, he photographed the architecture of Europe assiduously, and produced guidebook photographs of high quality. Everywhere he traveled, he went with his camera at the ready, so to speak.

Wachsmann's peregrinations through Europe took him, in September 1934, to Yugoslavia. At Spalato, standing camera in hand in a narrow gallery of a church tower, he fell into conversation with a fellow German tourist, about cameras and the art of photographing architecture. This tourist, it soon emerged, was none other than Walter Gropius. For Konrad Wachsmann, a firm believer in coincidence as the hand of providence at work, this encounter with Gropius was indeed to be a fateful meeting. As far as can be ascertained, they had, prior to this, and strangely enough, never met in Germany. Of course Wachsmann knew of Gropius, who was famous throughout Germany; but whether Gropius had heard of Wachsmann, whose star was yet to rise, is unlikely. There had in fact been one point of contact, in 1930, but it was a purely formal one and involved no direct meeting between the two men. The magazine *Bauwelt* had at that time conducted a competition for "das billige zeitgemässe Eigenhaus." Gropius and Martin Wagner were among the judges, and among the many prizes awarded were two to entries by Konrad Wachsmann: a one-family house and a row-house design,[16] both using the Christoph and Unmack construction system, as well as the services of their cost analysis department. It is probable that Gropius knew of Christoph and Unmack. He was remarkably well informed in all matters pertaining to prefabrication. He had moreover in his experimental house at Stuttgart, used *Lignatplatten* for the internal lining, and these cellulose-fiber boards, made to a Swiss patent, were handled in Germany exclusively by Christoph and Unmack.[17] And of course he knew Poelzig, Wachsmann's patron, and a director of Christoph and Unmack, through many professional associations and through intimate contact in Der Ring. But prior to the meeting at Spalato, he did not know Wachsmann.

At this meeting Gropius was on the first leg of a journey into voluntary exile. For years, even prior to the assumption of power by the Nazis, in 1933, Walter Gropius, "that elegant salon Bolshevist," as he was derided in the party organ, the *Völkischer Beobachter*,[18] had been under National Socialist attack. He was assailed as an acknowledged leader of the international Modern Movement, as the founder of the Bauhaus ("the cathedral of Marxism"), and as an important member of Der Ring, (that "Jewish-Bolshevist architectural organization").[19] After his association with Hirsch—perhaps in part because of it—he found it difficult to find work, and the copper houses were to be his last important executed project in prewar Germany.

Early in 1934 he was involved in what was to be his swan song, a minor event in his career: the organization, together with Joost Schmidt, of the nonferrous metals exhibit at the Deutsches Volk–Deutsches Arbeit exhibition in Berlin. His participation in this exhibition was something of a test case for his future professional standing in Germany, as well as for the future of modern architecture as whole. After the dissolution of the Bauhaus in July 1933, and the vicious attacks on all associated with it, or with

6.3
Konrad Wachsmann, row house, competition entry, Das Billige Zeitgemässe Eigenhaus, 1931; judges include Gropius and Wagner

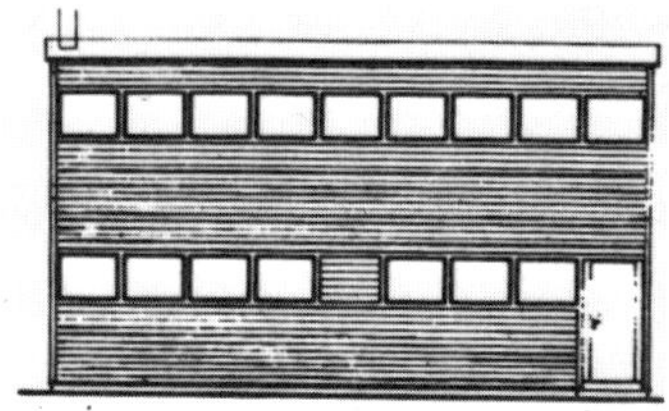

Straßenseite

Schaubild Gartenseite

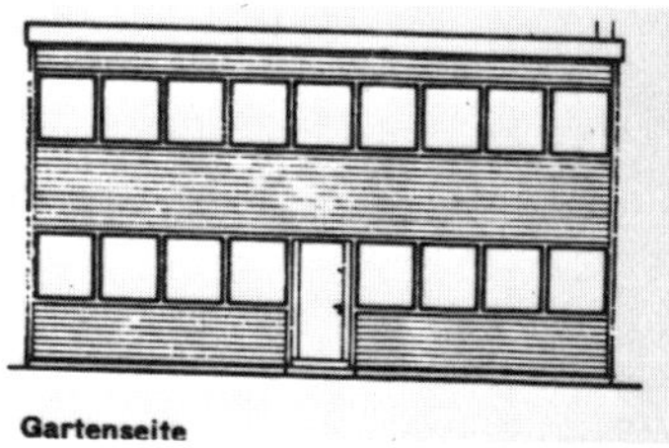

Gartenseite

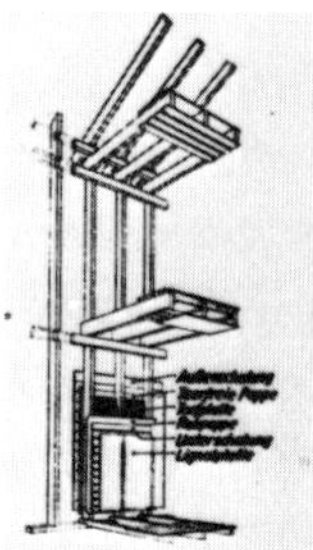

Wandkonstruktion

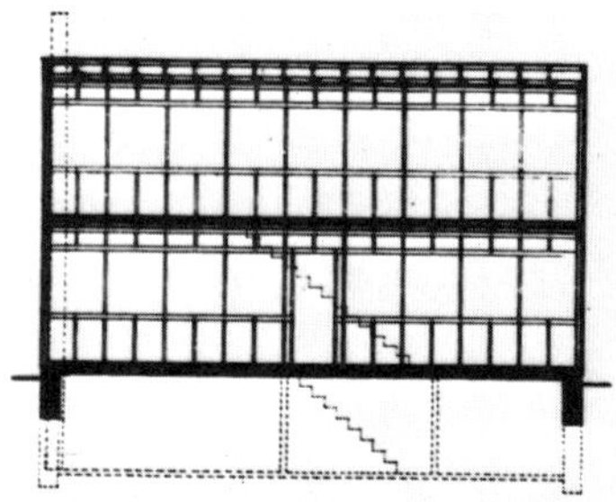

Schnitt

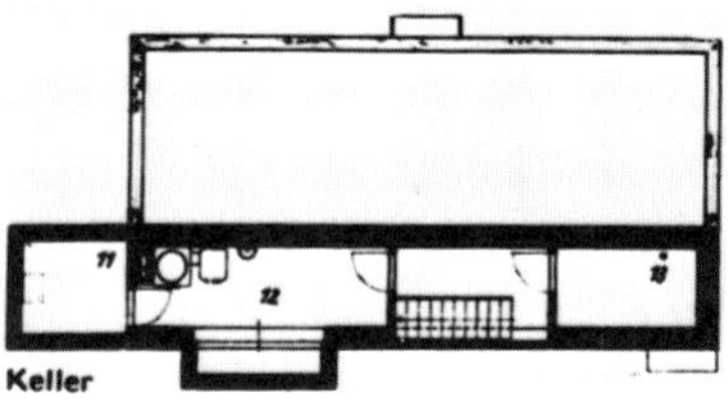

Keller

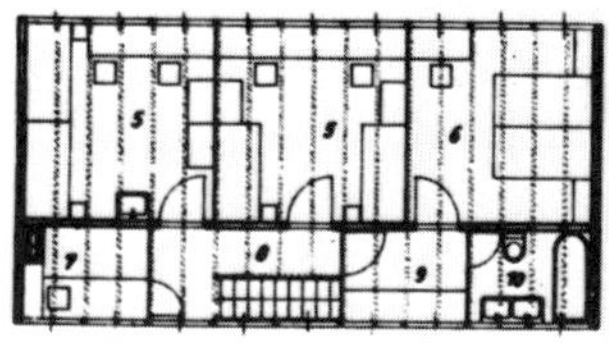

Obergeschoß
Maßstab 1 : 200

1 Speisezimmer
2 Wohnzimmer
3 Küche
4 Diele
5 Kinderschlafzimmer
6 Elternschlafzimmer
7 Kammer
8 Obere Diele
9 Schrankraum
10 Bad
11 Kohlen
12 Heizraum und Waschküche
13 Vorratsraum

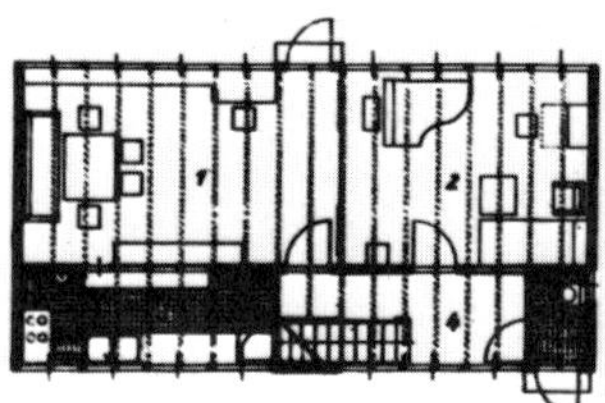

Erdgeschoß

avant-garde architecture generally, there was indeed much to fear. Gropius had made strenuous efforts to preserve the architectural movement to which he had dedicated his life's work, in letters to the new administration, in February and March 1934. His cry from the heart, "Can Germany afford to throw overboard the new architecture and its spiritual leaders, when there is nothing to replace them?"[20] fell, ominously, on deaf ears.

Something of the stresses of that period, but nothing of their true intent, was apparent to the eminent British architect Howard Robertson, who visited the Deutsches Volk–Deutsches Arbeit exhibition early in 1934. He reported home on his impressions of what he called "the first Nazi exhibition" in Berlin:

Two of the best Halls, those of chemistry and metals, show strong evidence of the influence of Walter Gropius and Mies von [sic] der Rohe. These modern architects, one is glad to learn, have not suffered total eclipse, in spite of the misfortunes of the Bauhaus. We are told that by official order they were not to be excluded from participation in the Exhibition. Their great international contribution to design has therefore been recognized; and this fact shows all absence of bigotry, which is reassuring in view of rumours of architectural reaction.[21]

What this "fact" showed, however, was not "all absence of bigotry," as was so sanguinely believed, but the effectiveness of Nazi propaganda in their effort to project a respectable image.

The harsh reality behind this facade was very different, as Gropius later recalled. "Right after that I encountered a rather disagreeable event; a uniformed Nazi patrol came to my house and told me I would resent [regret?] it if I let myself be seen again in a large exhibition called 'German

People, German Work,' in which I was building a good-sized department for the non-ferrous metal industry.''[22] Despite these threats behind the scenes, Gropius was too well known a personage on the international front to be openly persecuted by the Nazis: at least at that early stage when Hitler was still anxious for world approval. Therefore when an invitation came from the prestigious Royal Institute of British Architects for Gropius to present an exhibition of his work in London, permission was granted for him to leave Germany to do so. This exhibition was prepared in April 1934 and held from 15 to 26 May of that year. Consisting of ''170 drawings, photographs and diagrams illustrating Professor Gropius' work, ranging from the surprisingly modern pre-war buildings to recent planning and housing schemes,''[23] the exhibition was opened by Sir Raymond Unwin; it was an unexpected concession to modernism on the part of the conservative RIBA, which reflected credit on the intelligence of the secretary, Ian MacAllister, as well as on the initiative of Morton Shand, who actively promoted it.[24] It was also an indication of the respect in which Gropius was held, even in circles far removed from the front line of modern architecture.

At the time of the exhibition of his work Gropius had also given a lecture to the Design and Industries Association—the British offshoot of the Deutscher Werkbund—on ''The Formal and Technical Problems of Modern Architecture and Planning.''[25] Maxwell Fry, who had played a role in the exhibition, was in the chair at the lecture. ''I remember a meeting in the School of Hygiene and Tropical Medicine arranged by the Design and Industries Association, where Gropius introduced us to our more serious tasks in an atmosphere of tense excitement and breathless enthusiasm,'' he reminisced some 20 years or so after the event. ''I can remember exactly the overcrowded room and he standing among us, speaking with the utmost clarity in broken English. . . ."[26]

From this successful visit to England Gropius returned to a troubled Germany. It became increasingly apparent that he could not remain there much longer. As he conceded in London, he was "not exactly a *persona grata* with the Chauvinists" of Germany;[27] and he now looked for a way out. Through the mediation of Jack Pritchard, a patron of modern architecture and industrial design,[28] Maxwell Fry had suggested that Gropius be associated with him in two architectural projects in England. A carefully worded letter had been written to Gropius, which would be helpful in getting permission to leave. Now, in addition, an invitation was received from the Fondazione Alessandro Volta, for Gropius to participate in an international conference on the theater to be held in Rome from 8 to 14 October 1934. On the grounds of these international approaches, which could only bring prestige to Germany, the Nazis granted Walter and Ise Gropius a "leave of absence" permit[29] for what was presumably to be a temporary visit. For the Gropiuses this invitation was to be "both the pretext and the means to escape from Hitler's Germany."[30] On their way to Italy they stopped off in Yugoslavia, and at Spalato they met Konrad Wachsmann.

Gropius and Wachsmann: Encounter and Reunion

Between Gropius and Wachsmann, a quick friendship sprang up, as they discovered shared interests. They planned to meet again the following week, this time in Venice, and from there they went on together—Walter and Ise Gropius, and Konrad Wachsmann—to see the mosaics in Ravenna. Wachsmann, the eternal romantic, was overwhelmed by the magic of it all—the illustrious friends he had made, the wine, the gondolas, the music: "es war ein phantastische schöne venezianischer Tag."[31]

Eventually, a month later, they came together in Rome, where Wachsmann had set up in architectural practice and where Gropius, as a guest of the Italian government, was to address the Volta conference on the theater.[32] Between them they had very little money, Gropius with "ten marks in his pocket," as German currency controls were severe, and Wachsmann chronically short of funds. But Wachsmann had a small open car, and they spent an exuberant time together. Circumstances forced them into a certain intimacy of association. They were both out of their normal milieu, expatriates uncertain of the future; they had parallel architectural experiences in the past, a mutual professional interest in prefabrication; and despite their disparate personalities, both shared a great zest for life. It may have been temporary, like a shipboard friendship, but in the few hectic days and nights they shared in Venice and Rome, a favorable chemistry bound them together.

From Rome Gropius went on to London.[33] Jack Pritchard, who had previously commissioned the first modern block of flats in London, the Lawn Road flats designed by Wells Coates, allocated an apartment to the Gropiuses; and Gropius went to work in partnership with Maxwell Fry. After this promising beginning, life became increasingly difficult. Very few projects on which he and Fry worked in the next three years reached concrete realization—just the Impington College and a couple of houses—but although this was discouraging, Gropius' courage never flagged.[34] The possibility of a chair at Cambridge University was mooted but failed to materialize.[35] Most trying of all, the years passed, and what had at first been conceived as a short episode away from Germany, while things settled down, now appeared likely to be a long-term exile,[36] cut off from friends, family, and culture in a rapidly polarizing world.

In 1937, unfulfilled in London, Gropius responded to approaches made by President James B. Conant of Harvard University, and Joseph Hudnut, Dean of the Graduate School of Design, and accepted an offer of a professorship at Harvard.[37] Through the intervention of Ernst Jaeckh, onetime executive secretary of the Deutscher Werkbund—who had access to the Nazi administration[38]—Gropius received permission to return discreetly to Germany to retrieve his furniture, paintings, and other possessions, including all his documents, from the care of his sister-in-law, and to sail to the United States "without unpleasant remarks in the press"[39] or other, more drastic, sanctions. They arrived in Boston in March 1937, to begin a new life. Gropius rested for the summer and then took up his appointment in the Harvard architectural school. There he was soon joined by some of his former colleagues, including Marcel Breuer of the Bauhaus and Martin Wagner, his frequent associate of Berlin days.

Although not without economic and emotional stress, as the trauma of his uprooting from Germany was exacerbated by the growing threat of war, the next years were good for Gropius. He made a respected academic niche for himself, being appointed chairman of the architectural department; he built his own house in Lincoln, and for the first time in many years could feel secure, as he began to put down new roots. Building opportunities came to him, albeit slowly, as he began to establish himself in architectural practice in association with Breuer. His opinions were widely sought after, and if he did not always evoke agreement, he was always listened to with respect. When the war in Europe broke out, toward the end of 1939, he could confront its problems, especially as they affected his friends and family in Germany, at least from a secure and stable base. Wachsmann's situation, on the other hand, had by this time deteriorated seriously. After Gropius left Rome, at the end of 1934, Wachsmann had stayed on, to practice architecture there. It will be recalled that Wachsmann's architectural training had been somewhat desultory and that his experience in architectural practice had been almost exclusively confined to building in timber. This of course was hardly the appropriate basis for working in Italy. But Wachsmann had initiative, an eager and open mind, and an unerring sense of construction. He therefore turned with ease to the predominant building material of the new Rome and began to produce innovative structures in reinforced concrete: multistory apartment buildings, a covered market structure, and a multistory office building with underground cinema and parking garage. These were ambitious projects, few of which came to fruition, but they stretched his knowledge and his ability as an architect. In the normal course of events this could have been the modest beginning of a new and successful career. But these were hardly normal times, and Wachsmann could not escape the realities of Europe in those fateful years, as crisis followed crisis and the world scene steadily darkened. An increasingly militant Fascist Italy, now more strongly in alliance with Nazi Germany, no longer provided a safe haven for the German-Jewish exile. In 1938, therefore, he uprooted himself once more and moved to France.

Here, constantly harried and always poverty stricken, he moved anxiously from one temporary refuge to another. Accompanied by the woman who was later to become his wife,[40] he made his way from Marseilles to Aix-en-Provence, and from there to Paris. After several months they moved on to Grenoble where they had learned it was possible to acquire identity cards, which legitimized their stay in France. It was in Grenoble, inspired, as he said, by the slender steel lampposts of that city, that Wachsmann conceived the tubular steel space-frame construction system which formed the basis of his later, renowned Mobilar hangar. Despite help from friends, some fellow refugees, life was lived on a hand-to-mouth basis. Then an English poet, Harold Tooby, offered a place of refuge: a primitive stone-built sheepherder's hut on the Cote d'Azur between Vence and Les Tourette, with a "breathtaking view," which, together with the security if offered, helped compensate for its almost total lack of amenity. Here Wachsmann was active in trying to assure himself some kind of future. He sent out letters of appeal for help to influential friends now abroad, to Walter Gropius and to Albert Einstein. He also invested much effort in trying to get his space frame patented and commercially exploited, an act of incredible optimism considering his precarious position. And finally, he applied for immigration visas to the United States, with little hope of success as America pursued its tragic restrictive policies.

In miserable condition, in the minimal shelter of the hut near Vence, always penurious, usually hungry, they managed somehow to survive. Albert Einstein, who despite his eminence, had been deprived of his German research post and professional status, was now in honored exile at Princeton; notwithstanding the many other calls for help to which he responded, he sent for some time a monthly retainer of $10 to Wachsmann, an act of grace which meant "salvation" to his former architect. In the meantime the threatening clouds of the coming war grew ever darker in Europe. In September 1939 Hitler marched into Poland, and the Allies declared war on Germany. All aliens in France were immediately ordered to report to the police. Wachsmann was of course at this time entrapped in the eternal no-man's-land of the refugee: to the Germans he was a Jew and therefore an outcast, a nonperson; but to the French he was a German and therefore an enemy alien. As such he was immediately interned, first at Antibes, then in other camps. As a "carpenter and architect," as he styled himself, he soon found himself active, designing and constructing barracks for his fellow inmates.

At this critical juncture, early in 1940, Walter Gropius came to his aid, writing forceful letters to high French officials, trying to get Konrad Wachsmann, "a very good friend of mine," as he put it, "an extremely talented young architect . . . a most distinguished man of high standards," out of the *camp de rassemblement* in France.[41] In the spring of 1940, as a result of these efforts—Gropius was tireless in his concern for the well-being of his friends and former colleagues now trapped in Europe—Wachsmann was freed from the internment camp and almost immediately inducted into the French army. His military service was short-lived, however, and he was demobilized shortly thereafter, when France capitulated in June 1940 to the Nazis. He managed to get to Aix, was there reunited with his companion, and there they eked out an existence until May 1941, when on his 40th birthday, they eventually succeeded in getting visas to the United States. A few months later, after a long and circuitous wartime voyage, Konrad Wachsmann arrived, a destitute refugee, in New York. Gropius, who had been in frequent contact with the Emergency Rescue Committee in New York, offered to give Wachsmann accommodation in his own home for a while and sent him the rail fare to come to Lincoln. There, in the ambience of warmth, friendship, and hospitality traditional to the Gropius home, he found shelter and security after his long odyssey.

Wachsmann's arrival in the United States was timely, from many points of view. He had come almost at the last minute, while America was still a neutral country in a world at war, and there he found, in its general and more immediate sense, a climate propitious for the development of plans he had long cherished for the industrial prefabrication of the dwelling. Even when in France, after joining the army, and while in that capacity designing a house for an army doctor, Wachsmann had worked on such a proposal. ''Probably inspired by the barracks I had built,'' he recalled, ''I also laid down the principal idea for a mass-produced and industrialized housing project.'' In the French debacle such plans were intellectual exercises, therapeutic perhaps but hardly of practical import. Now, in the United States, the situation was vastly different. Industry in America was gearing up for war, and these preparations generated a demand for housing that was as large as it was urgent. A housing crisis was at hand, and it was in times of crisis that prefabrication came into its own. By late 1941 the American prefabrication industry at last appeared to stand on the threshold of maturity, ready to respond to the crisis in housing. If its past development had been sporadic and hesitant, its future, from Wachsmann's perspective, seemed more hopeful than he could have dared to dream.

7

The American Prefab as a Crisis Response

The Depression: Housing the "Other Half"

In 1932, Kenneth Kingsley Stowell, in an article entitled "Housing the Other Half,"[1] sought to draw the architectural profession's attention to the critical dimensions of the housing problem then being experienced in the United States. This was at the climax of the depression, when millions were jobless, and housing production had declined precipitately by 84 percent from the 1922–28 average.[2] To cope with this crisis, Stowell advocated, among other steps, the radical reorganization of the building industry, in order to exploit the potential of industrially produced housing; the technical problems involved, he believed, were on the way to being solved by the pioneer prefabricators. Stowell here highlights the two main themes which are constantly to be reiterated by protagonists of the factory-made house: the theme of the housing crisis and the potential solution of that crisis through the prefabrication of dwellings. Now we have pointed to the fact that the history of the development of prefabrication has very largely been a story of crisis and response, and it is true that the advances in prefabrication in Germany in the 1920s, in which Gropius and Wachsmann had been involved, were similarly made in response to the social and economic stresses of that era. But the problems they were addressing, with Hirsch and Christoph and Unmack, were not the basic housing problems of the working class, or—despite the "back-to-the-land" movement—of the unemployed. The prefabrication of the one-family house in Germany, although it was seen as a possible solution to certain middle-class housing problems in an age of eroding bourgeois standards, was nevertheless marginal, in quantitative terms, to the housing crisis of the poor. Gropius of course fully realized this. The proper response to the challenge of housing the urban poor, that is, of responding to the housing crisis in its most urgent sense, was seen in the social housing of the *Siedlungen* in such cities as Berlin and Frankfurt. These mass-housing projects were large-scale and comprehensively planned urban areas, usually on open land on the periphery of the larger cities. They involved a synthesis of advanced concepts of architecture and urban and social planning. In construction they were technically sophisticated and generated their own significant experiments in industrialization. However, although some of these, such as Gropius' rationalization of the building process at Toerten-Dessau or Ernst May's ambitious program of standardization and prefabrication at Frankfurt, advanced the cause of industrialized building, they were exceptions and not the norm. The considerable achievements of German social housing, indeed of European housing in general, were based not on total systems of prefabrication but rather on highly mechanized and efficient uses of conventional building methods, together with the extensive application of factory-made components.[3] We have the anomalous situation therefore of the technologically most advanced systems in Germany, the fully prefabricated systems in steel, timber, and copper, being directed to what was, in that social environment, a retrogressive market, while the socially radical housing of the *Siedlungen* was built largely by the traditional (but efficiently organized) building industry.

Now, in the United States, the concept of social housing was not conceived in these radical terms. Except for a few notable, and relatively small-scale experiments,[4] there was to be no American experience in the 1930s to parallel that of Germany. This we say notwithstanding the impetus to social housing given by the programs of the Reconstruction Finance Agency and the Housing Division of the Public Works Administration, in the early years of the decade, as a response to the crisis of the depression. Nor did the admiration for German models (and especially for the work of Gropius and May) expressed by those architects[5] involved in some of the pioneer efforts in the United States generate a wide movement to emulate the European achievement. Catherine Bauer summed up the housing situation in 1934, in these depressing terms:

Throughout the country, then, there are not more than twenty thousand dwellings erected since the war on a permanently nonspeculative basis, and with any pretensions to large-scale planning or fundamental change in the quality of house production and neighbourhood environment. Twenty thousand to set against 4,500,000 in a section of Europe with only slightly more population than that of the United States. Moreover, not more than half of the twenty thousand really achieve a degree of permanent amenity and freedom from congestion which is the minimum working standard for "modern housing" in Europe. And of the remaining ten thousand few or none were available to the lower-paid half of the population who need the houses the most.[6]

Housing the "other half"[7] was clearly not being achieved, least of all in terms of the European notion of comprehensive planning on a mass scale. The European concept of social housing, so alien to American perceptions and expectations, was thus not seriously proposed as a viable alternative, neither by the administrators—except on a petty scale—nor by the architects and builders responsible for solving the housing problem. Mass housing was perceived in the negative sense of "tenement building"; and the true dimensions of the housing problem were not adequately sensed. On the other hand, although the crisis generated no radical proposals in a social sense, it served sharply to focus attention on the accelerated production of the normal housing product, the one-family dwelling. Technical ingenuity, in conception and manufacture, became a preferred direction of development. Prefabrication seemed an answer to the problem, once it was conceived in technological rather than social terms, and as it was advocated in the early 1930s, in the United States, this was seen as the industrial production of the individual house as a product—to a greater or lesser degree of technical sophistication—analogous to the automobile. This, after all, was the universally accepted paradigm of the industrial process, a success story of ever-increasing production at ever-lower costs. The car was the obvious model for such advanced concepts as Buckminster Fuller's Dymaxion house of 1927, but it was also the underlying, if not apparent, presupposition of a large number of less radical, technically more conservative, systems of prefabrication, with which the pioneers of the twenties and thirties experimented. With the individual house the preferred solution to the need for a dwelling, the prefabrication of dwelling units seemed not only appropriate, but standing a real chance—as it had not done in Germany—of making a significant contribution to the housing problem. Mass production, rather than mass housing, seemed to be the answer to a desperate need. There was one basic if untested assumption held in common by all advocates of this housing strategy: the factory-made house would significantly lower housing costs. But the economic justification, however essential to the viability of the concept of prefabrication, was perhaps not its prime motivation. In the American context, this was, somewhat surprisingly, philosophical. In discussing the use of technology to "mass-produce tracts of one-family houses," Catherine Bauer claimed in retrospect, "we chose individualism rather than collectivism."[8] In the light of this philosophy it is not surprising that the initial impetus to the development of prefabrication in the United States came not from the constructors of large-scale buildings but from those sectors of the building industry much more directly concerned with constructing and marketing the individual house, or supplying the components and materials necessary for its manufacture. The role of big industry—the steel companies and the large manufacturing companies—was, as we shall see, indirect, ambiguous, and in the long run indecisive, in developing and supporting the movement for the prefabrication of the dwelling. Even the most ambitious proposals thought only of the production of the one-family dwelling or its components.

Growth of an Industry

By the early 1930s, within the parameters of this approach, there was some considerable ferment of ideas in the technological field, resulting in a fair amount of practical exploration. Spurred by the depression and its "confluence of factors, economic, social, and technical," this activity led to prefabrication becoming, in Burnham Kelly's phrase, "a widely recognized movement," gaining the interest of many beyond "the handful of inventors and small companies which had previously been concerned."[9] The dimensions of this new movement are indicated by the fact that, by 1935, some 33 prefabricated systems were being offered on the market.[10] It was, however, nearly a decade behind Europe, and on a much more tentative scale to begin with.

Although the conditions prevailing after the first world war had helped to stimulate the development of prefabrication first in Britain and eventually in Germany, that war had little direct effect on housing in the United States. The traditional building industry was flexible, and could expand its resources to deal adequately with all housing needs, even the accumulated shortage generated by the war. By 1921 construction activity in the United States was well above the prewar index, and by its peak year, 1925, it was 2.6 times as great.[11] The demand for less substantial houses—temporary dwellings, farm houses, housing for oil-field workers—was catered for not only by the traditional construction industry but by several long- and well-established manufacturing firms producing precut and sectional houses.[12] Involved in this business were mail-order houses like Montgomery Ward, and especially Sears Roebuck, which, through vigorous promotion and a determined sales policy, coupled with liberal financing and loan facilities, managed also to invade the traditional housing market. Sears' "Modern Homes" department flourished with considerable sales of precut homes "packages" in the boom years of the mid-twenties, only to decline dramatically and ultimately collapse, in the general disaster following the crash of twenty-nine.[13]

Precut and sectional houses helped introduce industrialized methods into house construction but were not prefabrication in the full meaning of the term. Neither could construction in precast concrete be so considered, although it could reduce the site-labor content, sometimes significantly. There had been a continuing interest in precast concrete systems, since the days of the pioneers: since Ransome produced his "Unit System," utilized mainly for industrial buildings, or Conzelman, with a system similarly named, developed between 1910 and 1916, constructed a group of 300 houses of reinforced concrete precast elements at the Industrial Housing Colony in Youngstown, Ohio.[14] That this was a concept acceptable to constructors was noted in *Engineering News* in 1916: "There is nothing new about the so-called Unit System of concrete construction. It is of interest to record that this system of construction appears to be coming into favor among engineers."[15] Large-scale, story-height precast panels, such as Grosvenor Atterbury had used for the Russell Sage Foundation at Long Island, or May at Stuttgart, were the logical form in which to prefabricate such elements, but they demanded the use of cranes and other mechanical hoisting devices, such as could be economically deployed only for extensive housing schemes on the European model. For the dispersed American market, with its emphasis on the individual house, this was clearly not appropriate. Many ingenious systems were designed employing smaller, lighter elements,[16] but these inevitably increased the amount of site work, especially in the placing of components and their jointing, usually by grouting. In addition in most precast concrete systems there was considerable on-site labor involved with internal partitions and finishes.

The movement for full prefabrication in the United States was reinforced in some measure through the European experience in the mid-twenties. The experiments carried out in Britain and Germany with steel housing, with its apparent use of high technology, and a sophisticated employment of materials superficially analogous to those used in the making of automobiles, attracted the interest of certain sectors of the American steel industry. A series of articles in the journal of the industry, *Iron Age,*[17] during 1926, drew attention to the British experience with the Weir, Telford, and Atholl steel houses, argued the technical and economic advantages of steel construction, and foresaw the potential for industrialization: "The development of standard, unit-type, interchangeable metal parts adaptable to any ordinary house plan will permit the use of large scale factory production."[18] The author's views on prefabrication were close to those of Gropius, when he went on to advocate "the development of standardized units for construction, rather than . . . uniform design or duplicate dwellings."[19] And we may note the congruence between the American philosophy of the individual, and Gropius' emphasis on the need for standardization and diversity. The unit steel house was conceptually attractive to the American, both technically and ideologically, and when the U.S. Department of Commerce issued its report on the production of steel houses in Germany—a report to which we have previously drawn attention—*Iron Age* noted "considerable interest" on the part of domestic steel producers.[20]

One development in particular that drew the attention of the metal industry, but evoked no immediate response in the architectural journals,[21] was the copper houses produced by Hirsch. Data on these houses have been collected by the Copper and Brass Research Association of New York, and descriptions of the prototype house, and many illustrations of it during erection and after completion, were published in the trade journals *Iron Age*, and the *Copper and Brass Bulletin*.[22] Attention was drawn, in both articles, to the advances made over the past decade in the design of what was called "knockdown" houses and pointed to Europe's lead over the United States in the design of substantial and permanent dwellings, that could be mass produced. That these could be expected to penetrate the American market was evident in the granting of an application for United States patents to Frigyes Förster. The message to Americans concerned with prefabrication was clear: metal construction was the technology of the future, and they were lagging far behind Europe in its development. Steel framing for houses of course had a long history in the United States, and the use of steel as a replacement for timber framing and studs goes back at least to a venerable example reported in *Iron Age* of 1907.[23] In fact Germany itself looked with some admiration to the American experience in this field, as we have already noted, and Spiegel, in his *Stahlhausbau,* devoted an entire chapter to the study of the framed structures pioneered by Robert Tappan, Weltcrete, Broderick and McKay,[24] generally in the mid-twenties.

Buckminster Fuller, in 1927, designed his revolutionary Dymaxion house, a lightweight metal structure suspended from a mast, which he exhibited in model form.[25] In its exploitation of advanced technology, and its departure from the constraints of traditional house forms, it was conceptually of an entirely new order, as compared with its European and contemporaries. In its original version it was obviously not ready for translation into practice, but later, development brought it closer to the stage of production. However, despite industrial backing,[26] optimistic prognoses in the press,[27] and Fuller's dedication and fervent advocacy, it never really got going: the Dymaxion house remained no more than a tantalizing promise, a highly innovative prototype of what an industrially produced metal house might be.[28]

It was only in 1932 that the first experimental steel houses were produced in the United States, using steel not only for framing but as the enclosing membrane in a total integrated building system, on the British and German models. "In recent years," it was pointed out in the *Architectural Record,* "vitreous enameled sheet steel has been used for the exterior surface of a number of residences in Europe. In the United States, however, the year 1932 appears to have seen the first attempt to utilize enameled sheet steel for this purpose. . . . The first house so covered was designed by Charles Bacon Rowley and Associates in Cleveland, Ohio."[29] There is a nice histori-

7.1
Robert Tappan, steel frame house, 1927

7.2
Buckminster Fuller, Dymaxion house, 1927

cal irony here. In 1932, as American architects and industrialists moved to produce steel houses, inspired largely by German models, Germany itself ceased their manufacture, to all intents and purposes. At that showcase of advanced prefabrication methods, the Wachsende Haus exhibition of 1932 in Berlin, there was not a single system utilizing metal-faced panels, other than the Hirsch houses designed by Gropius: and Hirsch themselves went out of business in 1932. Even Böhler, a pioneer of steel systems, had by 1932 moved to steel frame construction, with lightweight block or asbestos infill. A strange cycle of events was being played out. When, in 1926, prompted by over-capacity of production in a time of reduced demand, the German steel industry had decided to enter the steel house business, they were in part motivated by the British Weir house; they had failed, however, to study those problems to which the Weir house was about to succumb. Now, in the depths of the depression, American industry of diverse kinds, was similarly motivated to increase its sadly reduced production by entering the vast, unsatisfied housing market. And they too turned to the steel house as their paradigm, its virtues having been much more widely publicized than its faults. As Santayana once said, He who does not learn from history is doomed to repeat it.

By 1932 a frameless steel house using Armco steel panels[30] had been erected, on lines very similar to those of the pioneer house of Rowley, by the American Rolling Mills, "the first big industrial concern to eye the new market with serious intent."[31] In the same year two major firms were founded: Howard T. Fisher's General Houses Inc.,[32] which produced an all-steel house assembled out of components made by a large number of contributing firms, but undertaking no manufacturing on its own account; and architects Holden and McLaughlin's American

Houses, Inc.,[33] which manufactured a steel-framed, asbestos-clad product. These two corporations were, for a short time, managed by a holding company, Houses, Inc., formed by Foster Gunnison, in collaboration with the American Radiator Co. (a prime backer of the important research institute for prefabrication, the Pierce Foundation) and the General Electric Company, until Gunnison set up his own firm to manufacture stressed skin plywood houses, in 1935.[34] These experiments with steel houses were described in the *Architectural Forum* with interest, but not with undue commitment or faith in their future: "The makers of prefabricated steel houses have gone ahead preparing a product which the public certainly does not want right now, may not want for many years."[35] Despite this professional caution, and "the competition of subdividers and the deeply entrenched likes and dislikes of the American Public,"[36] the newspapers seized upon the new development and, through widespread publicity, generated much interest, and perhaps an inflation of anticipation for what the new industry was likely to achieve. When General Houses was incorporated, it was probably reasonable—after all, the name of the company was a conscious analogy of General Motors—for the *New York Times* to announce "General Houses, Inc. Formed to Market Ready-Made Steel Homes Like Automobiles" (although the heading " 'Fordized' Housing Plan of New Group" stressed the wrong make of car!). However, it was certainly over-optimistic to write of "Mass Production of Homes in View," and to suggest that prefabrication could lift the country back to prosperity, as did the *Times* in 1933.[37] However, by 1933, the mood of the country was slowly moving back to optimism: in March 1933 Franklin D. Roosevelt was inaugurated as president, and he proceeded vigorously to heal the economy, and restore confidence in the nation.

7.3
General Houses, Incorporated, prefabricated houses, c. 1933. *Left*: standard house types; *right*: standard interchangeable wall units.

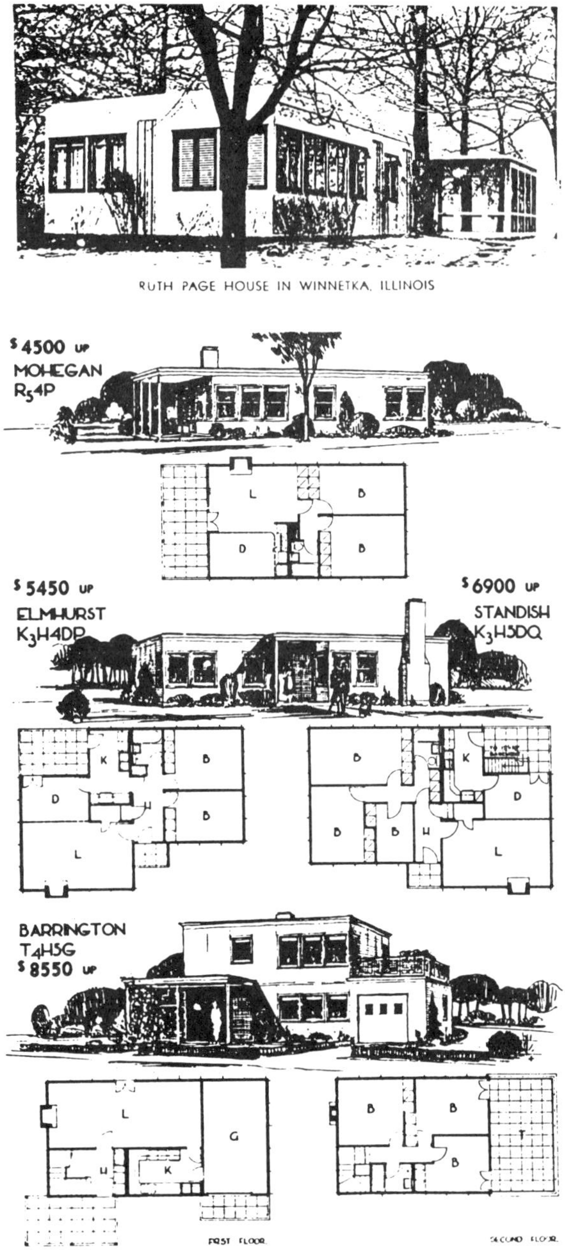

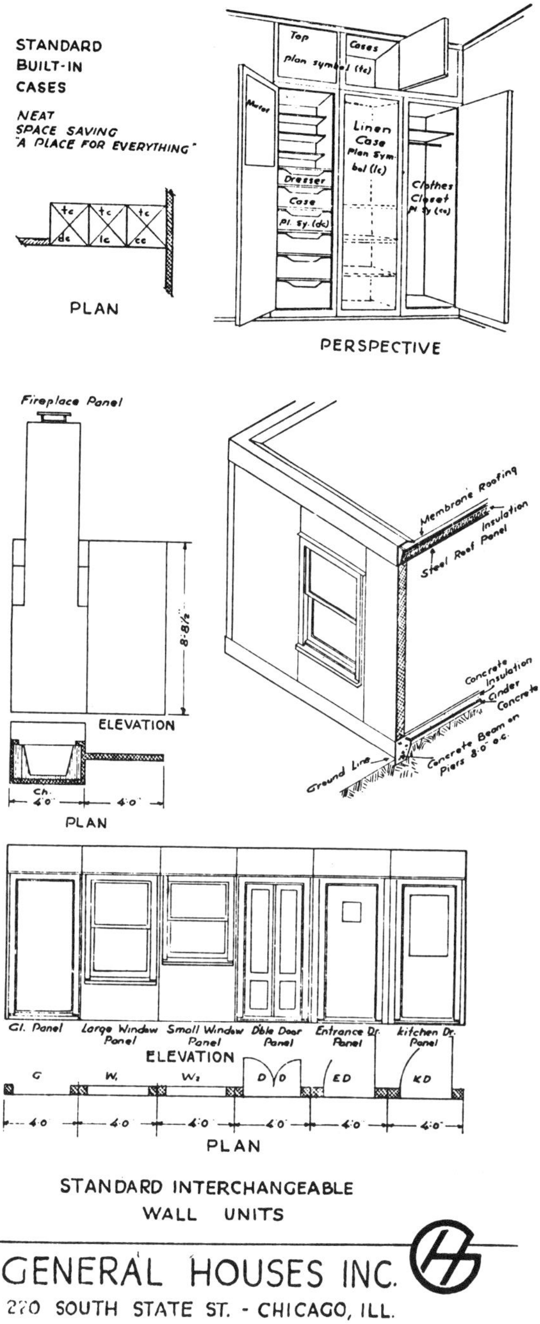
STANDARD
BUILT-IN
CASES
NEAT
SPACE SAVING
"A PLACE FOR EVERYTHING"
tc
dc
lc
cc
PLAN
Top
Cases
Plan symbol (tc)
Linen
Case
Plan Sym-
bol (lc)
Dresser
Case
Pl. Sy. (dc)
Clothes
Closet
Pl. Sy. (cc)
PERSPECTIVE
Fireplace Panel
8'-8½"
ELEVATION
Ch.
4'0"
4'0"
PLAN
Membrane Roofing
Insulation
Steel Roof Panel
Concrete
Insulation
Cinder
Concrete
Concrete Beam on
Piers 8'0" o.c.
Ground Line
Gl. Panel
Large Window Panel
Small Window Panel
Dble Door Panel
Entrance Dr. Panel
kitchen Dr. Panel
ELEVATION
G
W₁
W₂
D D
ED
KD
4'0"
4'0"
4'0"
4'0"
4'0"
4'0"
PLAN
STANDARD INTERCHANGEABLE
WALL UNITS
GENERAL HOUSES INC.
220 SOUTH STATE ST. - CHICAGO, ILL.

The Century of Progress Exposition held in Chicago in 1933 took place at a turning point in American history. The exhibition could therefore not only look back with pride to the achievements of the past but forward to the future with a confidence, shaken it is true by the depression but restored to a degree by the groundswell of hope kindled by Roosevelt's election. At the exhibition about fifteen modern houses were on show. Of these, several steel houses, by General Houses, Armco, and Stran-Steel, and Architect John Moore's wooden house of bolted panel construction, together with the exotic steel-frame and phenoloid board "House of Tomorrow" by George Fred Keck, were "the first of the prefabricated or mass-production types which the public has had the opportunity of investigating."[38] The architectural profession at large also began to learn about the new systems, both directly through the exhibition, or, in their offices, through the widespread and detailed coverage given to prefabrication in the professional journals.[39] Public and professional interest helped to give the expanding industry a firmer base; and it was also underpinned, in a technical sense, by the active programs of research into new methods being undertaken by the Pierce and Bemis Foundations, the Forests Products research laboratory, and Purdue University.[40] Also to be mentioned was the innovative program[41] of the Tennessee Valley Authority—part of the Roosevelt New Deal—which provided an early spark for what probably was the most successful (from a quantitative point of view) branch of the prefabrication industry, the mobile home. During 1935 and 1936 several new firms of prefabricators were established, and it is instructive to note that by now the preponderance of interest was no longer in steel but in the more prosaic, and better understood, field of timber construction, which could be operated with less capital investment. By the time Walter Gropius came to the United States, in 1937, there was not only significant activity but a climate of interest in prefabrication which was supportive of further work in this field.

The underlying factors of the housing situation were naturally not static. Economic revival was underway in what was becoming the postdepression era. In 1937, in his second-term inaugural address, President Roosevelt made housing his own policy concern, with the poignant phrase, "I see one-third of the nation ill-housed, ill-clad, ill-nourished"; and Senator Wagner at last achieved the enactment of the United States Housing Act.[42] Positive motivation, a new legislative framework, and decisive action achieved results. "By 1939, starts of new dwellings had recovered to about the 1929 level but with the help of 56,000 publicly financed units which were unheard of in 1929."[43] Despite these achievements, the housing problem, particularly of the low-income group, remained severe. Prefabrication of dwellings continued to be proposed as a possible solution.

Gropius drew attention to this in an interview published in the *Boston Herald*.[44] "There is something wrong if nothing on the housing market can be delivered to the ordinary man," he is quoted as saying. After referring to the high costs of land, he went on: "Mechanization of industry has brought down the costs of manufactured goods, but houses are still largely constructed by hand. One solution is the house made of parts constructed beforehand, the pre-fabricated house. These parts would come out of mass production—and there is no limit on the size of the building into which they would go. Of course, there are more than 50 American manufacturers dealing in prefabricated houses, but their enthusiasm has been slowed down because of the financial market and problems of land costs." He then concluded optimistically: "Changes for the better will come. Remember, food and clothing

7.4
George Fred Keck, "House of Tomorrow," Century of Progress Exposition, Chicago, 1933

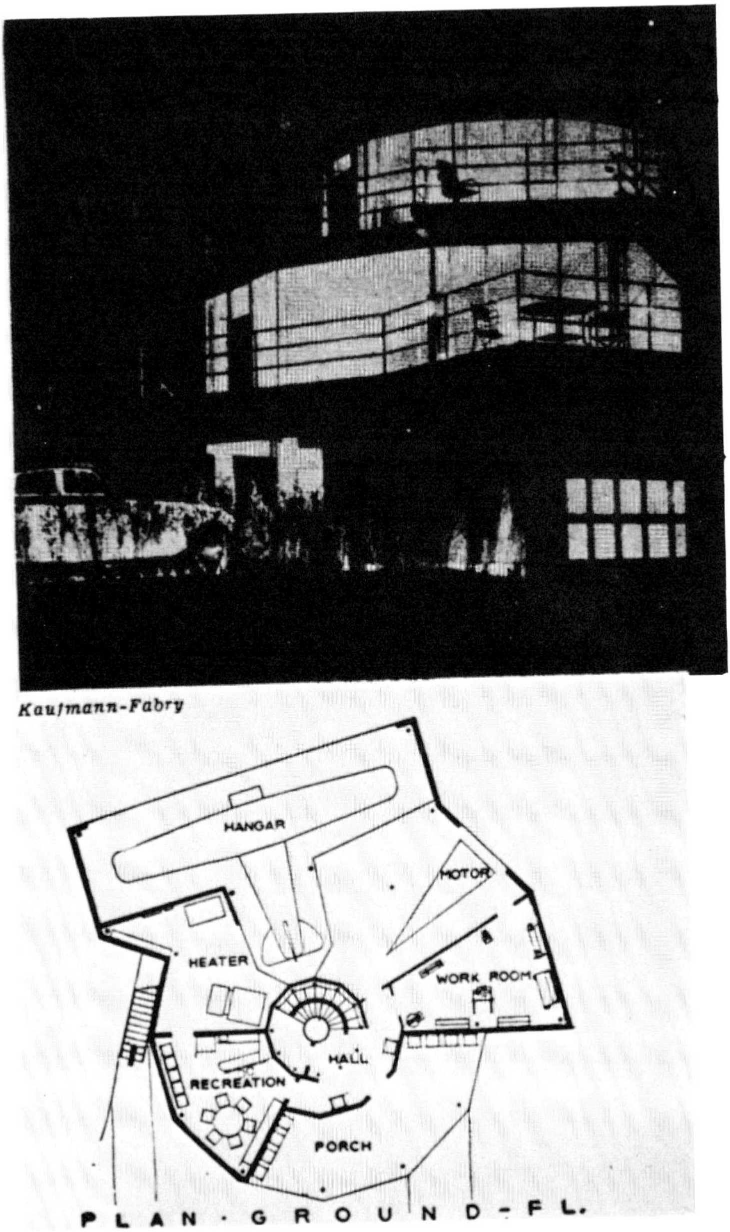

7.5
Forest Products Laboratory, stressed-skin plywood panel experimental house, 1935–37

7.6
Tennessee Valley Authority, truckable houses, c. 1941

prices have been adapted to the needs of the ordinary man, but not rentals. Some day they will be."

Gropius here compares housing with other commodities, food and clothing. The more usual analogy, which Gropius himself had cited many times and refers to again in the interview, was the automobile. Gropius saw the automobile as an example of industry's potential, but not as a model, for he was interested in producing components rather than the house as a complete product.

As Gropius was painting this somewhat hopeful picture of industrial effectiveness, the shadows lengthened on the international scene. By the end of the year Europe was once more at war, and although the United States remained, as yet, outside the conflict, the effect on its economic situation, and especially on its housing needs, was immediately felt. "The transition to a wartime economy in the United States began in 1940, even though under the pseudonym of 'defense' . . . . On June 28, 1940, the United States Housing Act of 1937 was amended to authorize the use of its loan and subsidy provisions for housing defense workers during the emergency . . . . On July 21, 1940, an Office of Defense Housing Co-ordinator was established in the Council of National Defense to plan and carry out defense housing programs. On September 9, 1940, $100 million was appropriated for the erection of defense housing by the War and Navy Departments. On October 14, 1940, the so-called Lanham Act, the basic defense housing law, was passed, involving direct Federal financing and construction."[45]

Walter Gropius and Martin Wagner

In this rapidly evolving new situation Walter Gropius played only a limited part. He continued to speak out for prefabrication, presenting the arguments, both pro and con, rationally. He spoke, not as a crusader for a cause but pragmatically. His measured terms of advocacy derived no doubt from his first-hand experience, in the Hirsch saga, with the enormous difficulties that stood in the way of realizing the dream of prefabrication. This caution provoked his old friend of Berlin days, fellow émigré and present Harvard colleague, Martin Wagner, to take him to task for his negative attitude, for stressing "the vast amount of problems concerned" rather than the great opportunities offered by prefabrication.[46] Wagner had an impressive record, in the city planning office of Berlin, of housing the masses. In 1931–32 he had demonstrated his interest in prefabrication by the guiding hand he gave to the organization of the Wachsende Haus competition and exhibition, and his book on the subject showed a firm grasp of the social and economic questions, as well as the technological aspects involved.[47] But his views on prefabrication were untempered by the flame of experience, especially in the new socioeconomic environment in which he now found himself transplanted, and they remained essentially academic.

Wagner's theoretical position on prefabrication remained ideologically consistent. He spelled it out in the manuscript of a proposed book, *Prefabricated Housing*,[48] on which he had been working for two years, since 1939. It owes much to the introductory chapters which he wrote for *Das Wachsende Haus*. His case for prefabrication grew out of a continuing and broad-based study of the housing problem (once again acute, because of the war situation). It is probably the most comprehensive study by an architect of its day, dealing with the socioeconomic dimensions of the housing question; the need for a new approach to construction management; and the technology of structure, enclosure, and environmental control. It is based on a clearly articulated thesis: that the proper module of industrialized housing, from both the sociological and technical points of view, is not the whole dwelling (vide Fuller), nor the component panel (vide Gropius or Bartning),[49] but the individual room. These rooms he sees as standardized cells linked together in free combinations to form an infinite variety of house plans. This inherent variability is an essential aspect of his philosophy of flexibility in house design, for which he makes an impressive case. That he argues convincingly for the "growing and shrinking house" comes as no surprise, when we recall his key role in the episode of the Wachsende Haus in Berlin. However, he is equally effective in dealing with other facets of dwelling flexibility, such as mobility and adaptability in the face of changing site and programmatic demands. After the breadth of this analytic presentation, his own solution to the problem (the "igloo house")[50] is idiosyncratic, but it should not be allowed to detract from the importance of his contribution as a whole.

Gropius and Wagner had frequently been associated in the past in various ventures: architectural competitions, housing schemes in Berlin, the inner circle of Der Ring. There was mutual respect, because between the two men's philosophy—despite the temperamental friction—there existed a basic congruence. Consequently for a while they were able to continue to work together and produced several studies in association.[51] Perhaps the most important outcome of this collaboration was a joint paper, "How to Bring Forth an Ideal Solution of the Defense Housing Problem?" which was presented as evidence to the House Select Committee investigating National Migration, of the 77th Congress, in 1941.[52] In this paper they advocated a program of prefabricated housing within the context of a long-term regional plan of decentralization. They also stressed the need for continuing governmental support of systematic research. Their

. . . A FRESH APPROACH TO HOUSING

leads to steel prefabrication, flexible size and igloo-like design. Martin Wagner's house offers protection against most everything, including air raids.

Basic unit of the MW house is a thirteen-sided conoidal room to which a small lavatory and stall shower are attached (below). Interconnected with small halls, these units produce a four-room house (above). "Murphy beds" drop down into the kitchen-living room or into the side-yards for outdoor sleeping. For a less radical variation of the basic plan, see top of next page.

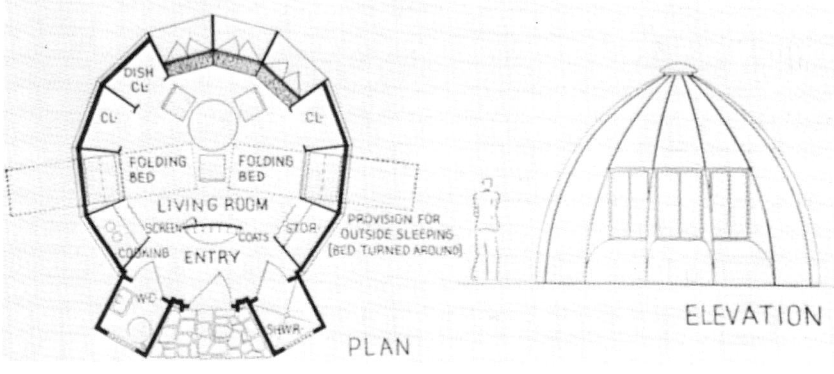

7.7
Martin Wagner, "Igloo" house, 1941

specific recommendation for prefabrication reads as follows:

To create a new type of low-cost dwelling of high quality, with up-to-date amenities and composed of standardized parts which should be interchangeable for use in different types of houses of varying sizes. These dwellings should be demountable for reerection but simultaneously to be qualified for permanent use when desired.[53]

We notice that the generic phrase "standardized parts" leaves open the disputed question, whether they are Wagner's room units or the panel components favored by Gropius.

In the published version of the proceedings of the Select Committee, the following item is noted in relation to the Lanham Act (passed to expedite the provision of housing in connection with national defense, and for other purposes): "At the time this hearing goes to press, a bill authorizing a further appropriation of $300,000,000 is proposed, making a total of $600,000,000 which will be available under the provisions of the act . . . ."[54] The effect of such massive underwriting of the housing program upon the prefabrication industry was noted in the *Architectural Forum*. "THE FORUM in December 1940 reasoned that 'national defense may do for Prefabrication what World War I did for the aircraft industry—raise it from infancy to adolescence in no time.' Since then Prefabrication has grown in stature, has grown from a group of experiments into a fledgling industry-within-industry which has been assigned some 17,500 houses for the Government defense program."[55] Gropius and Wagner, in the American context, were ideologues of prefabrication, and not practitioners (although Wagner's igloo house brought him into the realm of the "experimenter"). A month after this exuberant notice was published, Konrad Wachsmann arrived in the United States. With his arrival, he and Gropius took their first steps (with which Wagner was not associated) to enter the fledgling industry of prefabricators.

Part III

Dream and Reality: America in War and Peace

The turning point in building, reached long ago, does not mark a Renaissance or an ideological or spiritual reinterpretation of life; it is rather the culmination of the pressure, exerted by certain time-dependent, predetermined causes, of which it is meaningless to ask whether they should be accepted or denied. These are the new possibilities which society cannot afford to ignore; it is now our task to distinguish and comprehend them in all their potency and thoroughly master them, as the supreme instrument of creative activity. We have a long road ahead. . . . Industrialization is not a toy or a passing fashion . . . it is a building tool which must be mastered before any significant statement can be made.

Konrad Wachsmann, The Turning Point of Building, *1961*

For whatever profession, your inner devotion to the tasks you have set yourself must be so deep that you can never be deflected from your aim. However often the thread may be torn out of your hands, you must develop enough patience to wind it up again and again. Act as if you were going to live forever, and cast your plans way ahead. By this I mean that you must feel responsible without time limitation, and the consideration whether you may or may not be around to see the results should never enter your thoughts. If your contribution has been vital, there will always be somebody to pick up where you left off, and that will be your claim to immortality.

Walter Gropius, letter to a group of students, 1964

8

The Packaged House: A Wartime Proposal

The French Legacy of Konrad Wachsmann

When Konrad Wachsmann arrived at the Gropius house in Lincoln, Massachusetts, in September 1941, a destitute refugee, among his few possessions were two precious rolls of drawings, which he believed would one day make his fortune. One of these was the design of a tubular steel structural system.[1] This system, later to become widely known as the Mobilar hangar, was one of Wachsmann's major contributions to the art and science of building technology; its details fall beyond the scope of this study, but its development overlaps with our present concern and in some respects impinges seriously on it.

The second roll of drawings contained ten[2] small sheets, unannotated, unsigned, and undated, which delineated with exquisite precision a modular universal building system, consisting of load-bearing panels, weatherboarded externally, flush-paneled internally, thermally insulated, and combining freely (as indicated by the plans, sections, elevations, and details) to generate a house plan adhering to a rectilinear three-dimensional modular grid. The edges of the wall panels were beveled at 45 degrees, and were secured to each other by elaborate Y-shaped metal connectors. This proposal for a universal housing system lies at the heart of our subject.

8.1
Konrad Wachsmann, prefab system (French scheme), combination of panels, c. 1939

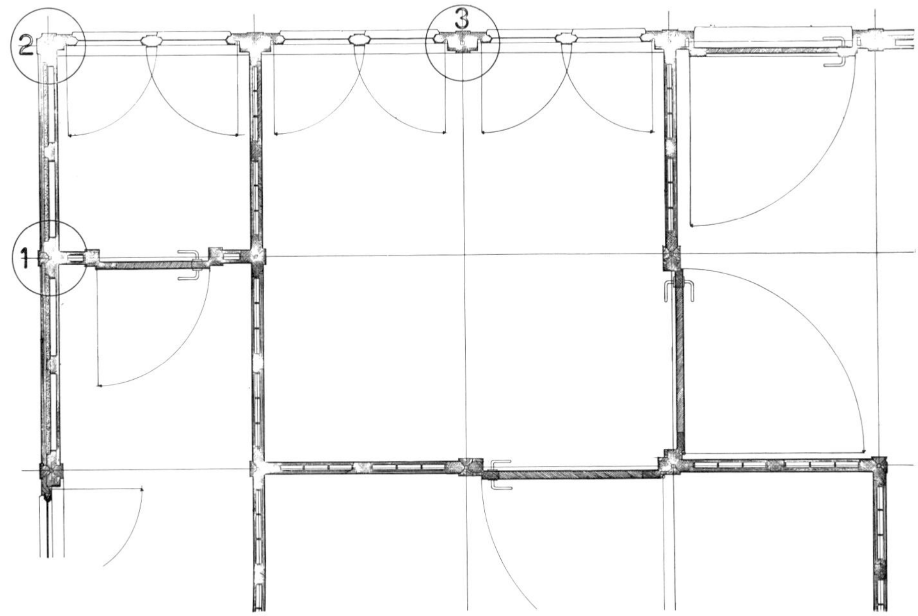

8.2
Konrad Wachsmann, prefab system (French scheme), metal connector, c. 1939

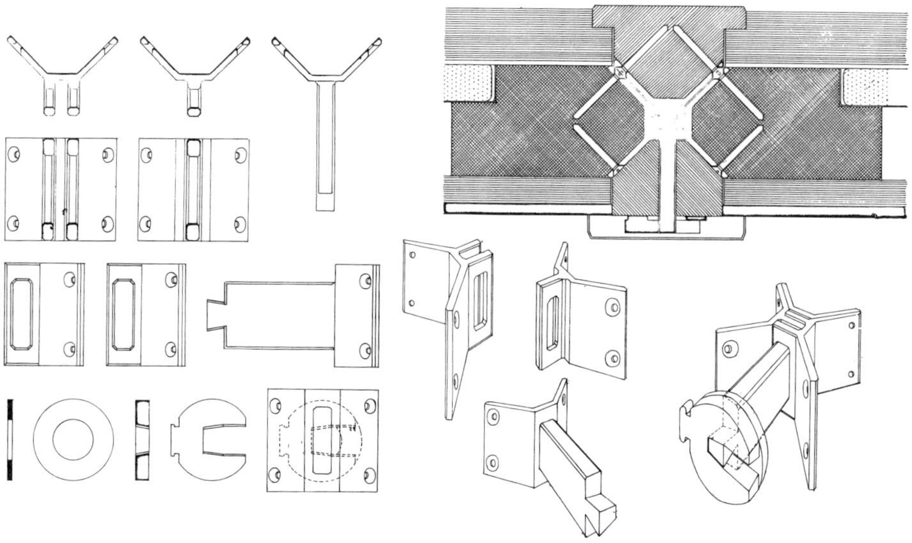

These two inventions, which represented the tangible legacy of Wachsmann's unhappy stay in France, were documented under the most adverse of conditions, the prefabricated house in the internment camp, the steel system partly in Grenoble and partly when sheltering in the South of France, in a "cave near Vence," in 1939.[3] On starting his new life in America, as Walter and Ise Gropius' guest, Wachsmann set these projects aside, to deal with more pressing immediate issues.
Seeking to give Wachsmann some source of income and, more important, a sense of independence, Gropius offered him the opportunity of working in association on two architectural projects for whose design he had just been commissioned: a recreation center for Key West, Florida, and a house for "a successful writer."[4] This new professional association was in a sense timely for Gropius, who rarely, if ever, put pencil to drafting paper, who always preferred to work in collaboration, and who, at the time of Wachsmann's arrival, found himself unexpectedly working alone, as his long-standing association with Marcel Breuer abruptly came to an end.[5]

Gropius and Wachsmann worked on these projects for a couple of months in the Gropius office in Cambridge, until the bombing of Pearl Harbor brought America into the war and eventually caused both schemes to be abandoned. However, the changed situation of America, and the implications of its new, active role in the war, stimulated Wachsmann to consider new architectural challenges and possibilities through the reactivation of dormant, but not forgotten, ideas. "That evening on December 7, 1941," he recalled, "returning home, I told Gropius for the first time that I had developed during the time in the internment camp in France a universal system of industrialized building components, of course in the metric system . . . we talked after dinner until late in the

8.3
Konrad Wachsmann, Ise and Walter Gropius, group photo, 1947

night about it. . . ."[6] The revelation of this system could not but excite Walter Gropius. He had, we may recall, just given evidence to the Congress on the need to develop such a factory-produced system of "standardized parts which should be interchangeable for use in different types of houses."[7] Now the basis of such a scheme, well conceived and presented, was put before him. The late-night discussion, on this fateful evening, could have but one result: a decision by Walter Gropius and Konrad Wachsmann to devote all their energies in a concerted effort to develop the proposal to its full potentiality. Gropius of course had many other obligations, by which his direct commitment was limited, but Wachsmann was free to give the project his full attention. To this task he now devoted himself, in a design studio set up in the basement of the Gropius house.

The Packaged House

Work now proceeded at a furious pace. Three drawing tables were set up in U-fashion, and Wachsmann, on a swivel chair at their center, worked day and night, seven days a week, hardly ever leaving the house. The first task was to convert the original drawings, done in the metric scale, to feet and inches. Wachsmann, who was never satisfied merely to reiterate a previous proposal, took the opportunity to reconsider many of the original details, most significantly the metal joint. This new set of drawings included details of ten standard panels which made up the set of enclosing elements, and a sectional perspective to show how they combined to create an architectural entity. It comprised 20 sheets, bound in a cover labeled "Konrad Wachsmann, 1941. 1–20" and constituted the definitive basis of all future cooperative effort.[8]

But development work did not stop there. On the contrary, yet another variant was painstakingly evolved, with a completely new type of metal wedge connector, the third which Wachsmann had devised so far. This wedge connector consisted of an interlocking set of metal plates housed in the panel edge, replacing the Y-shaped connector screwed to the beveled surfaces of the original and modified French schemes. As all the components of this new connector were essentially two-dimensional (rather than the complex three-dimensional form of the Y-connector), they were obviously considered to be easier to manufacture and less vulnerable to damage. Progress was unbelievably fast. At the beginning of February Wachsmann was able to report to a friend in England that the work was nearly done, with 20 out of 24 planned sheets completed.[9] He had worked compulsively, driven not only by his creative demon and his boundless faith in the system but equally by powerful emotional stresses of a blacker hue. He was an uprooted and displaced person, living on the kindness and hospitality of others; his wife Anna was ill in New York, and he was torn by anguish over the news from Europe that his mother and sister had been transported by the Nazis to Poland.[10] In his case work was not only a means of fulfilling long-held ambitions, and a way of regaining independence and self-respect, it was also an anodyne to pain.

The new set of drawings, comprising details of the panels, the new wedge connector, methods of jointing, floor and roof construction, stair details, room combinations, and illustrations of a "fictitious" building using all the elements, was completed by the third week of February 1942. Joseph Hudnut, Dean of the Graduate School of Design at Harvard University, was one of the first to see the completed work,[11] and suggested the name, the "Packaged House," which was immediately adopted.[12] The term itself was not new,[13] but it was appropriate and became the registered trade mark of the system as well as its popular appellation.

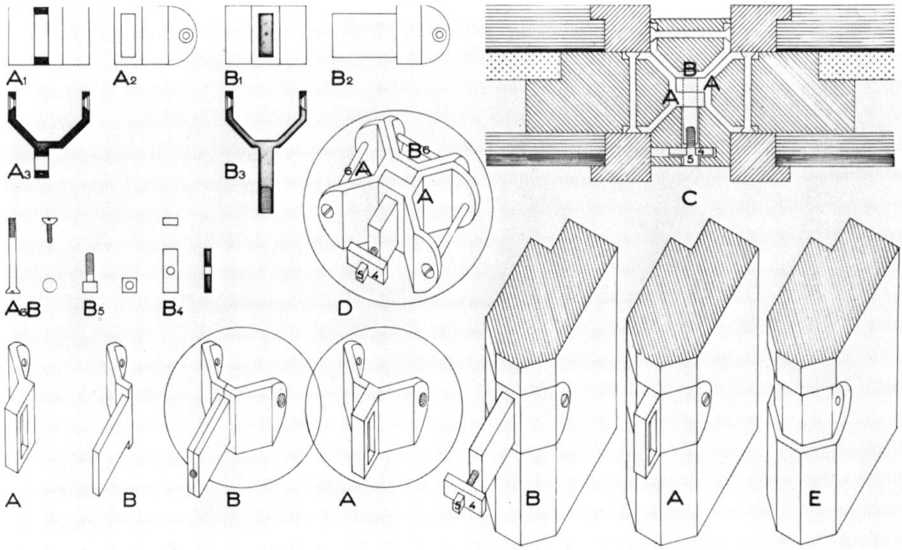

8.4
Konrad Wachsmann, modified scheme, detail of metal connector, 1941

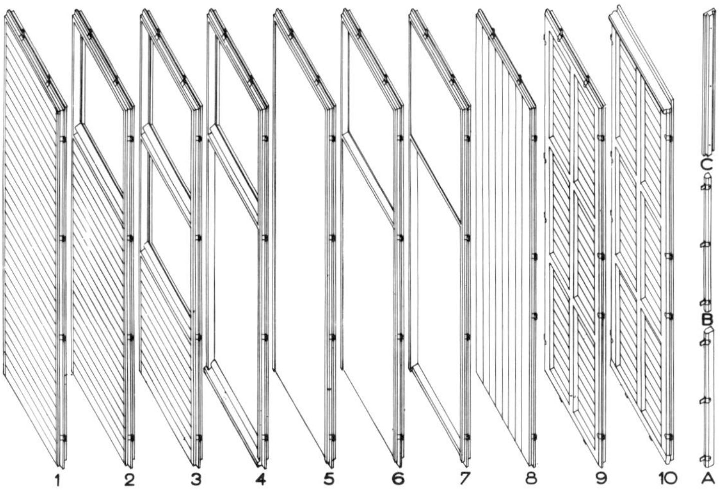

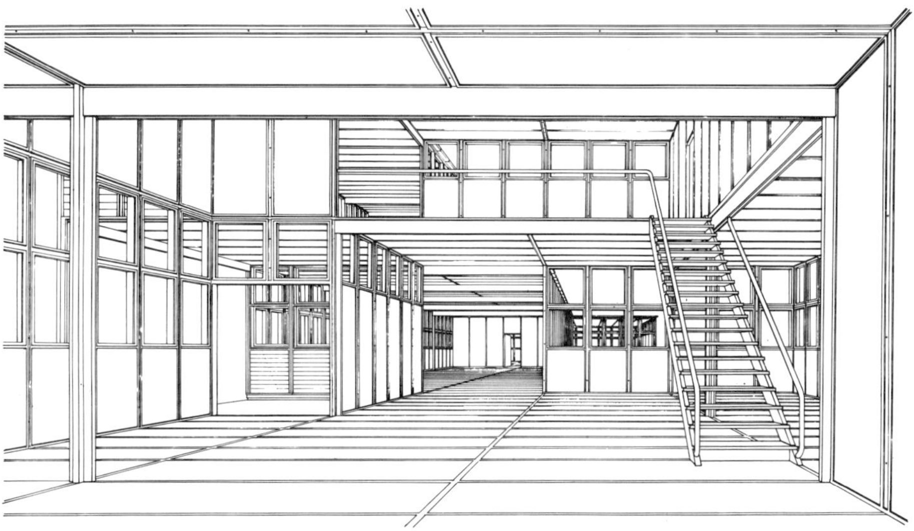

8.5
Konrad Wachsmann, modified scheme, panels, 1941

8.6
Konrad Wachsmann, modified scheme, sectional perspective, 1941

8.7
Wachsmann and Gropius, Packaged House, wedge connector, 1942

8.8
Wachsmann and Gropius, Packaged House system, 1942

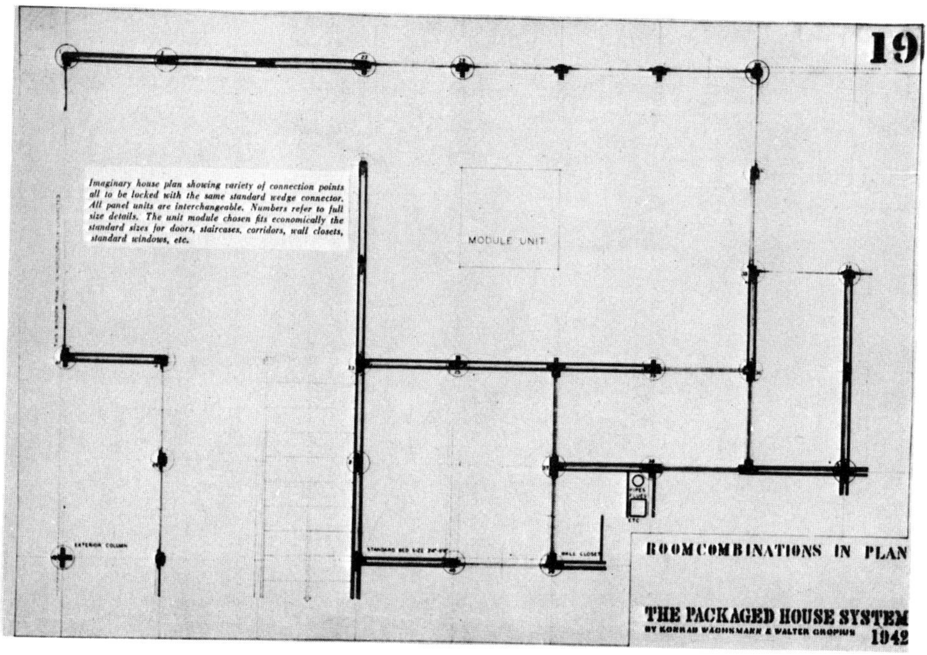

Gropius undertook the formalities of having the system patented, and by May 1942 an application had already been filed.[14] Although the patent application claimed the novelty of the connector as the fundamental innovation of the system, the points stressed in the objectives of the system were much more comprehensive: "The invention aims to transfer most of the labor involved in the construction of a building from the site of the building itself to a factory and to make the erection of the building primarily one of assembly." In order to achieve this, "standard units or sections, each consisting fundamentally of a duplicate of the other" are used so that "any frame section can be interchanged with any other." In this sense the system was conceived as universal, with an infinite potential for combination of a set of standard panels (the vertical ones being load bearing) related to each other in three directions.

In trying to place the Packaged House within the conceptual framework of systems theory, it is advisable to consider separately two different issues: the Packaged House as a construction system and the Packaged House as a design system. In construction terms it was conceived as a closed system; that is, it was entirely self-contained and comprised a kit of parts whose every component element was of necessity purpose-made. It was a rational system in that the number of components was severely limited; but for all that, it was closed. It did not seek to exploit the wide range of industrially produced building components then on the market; it could not, for instance, because of the nature of the panels and the jointing system, readily incorporate standard doors and windows, or pre-made roof trusses or ceiling panels, then being mass-manufactured by others. Nor did it adjust to industrial norms, where these existed, such as the standard 48″ width of plywood which was incompatible with its 3′4″ module. It was a closed, rigid, homogeneous,

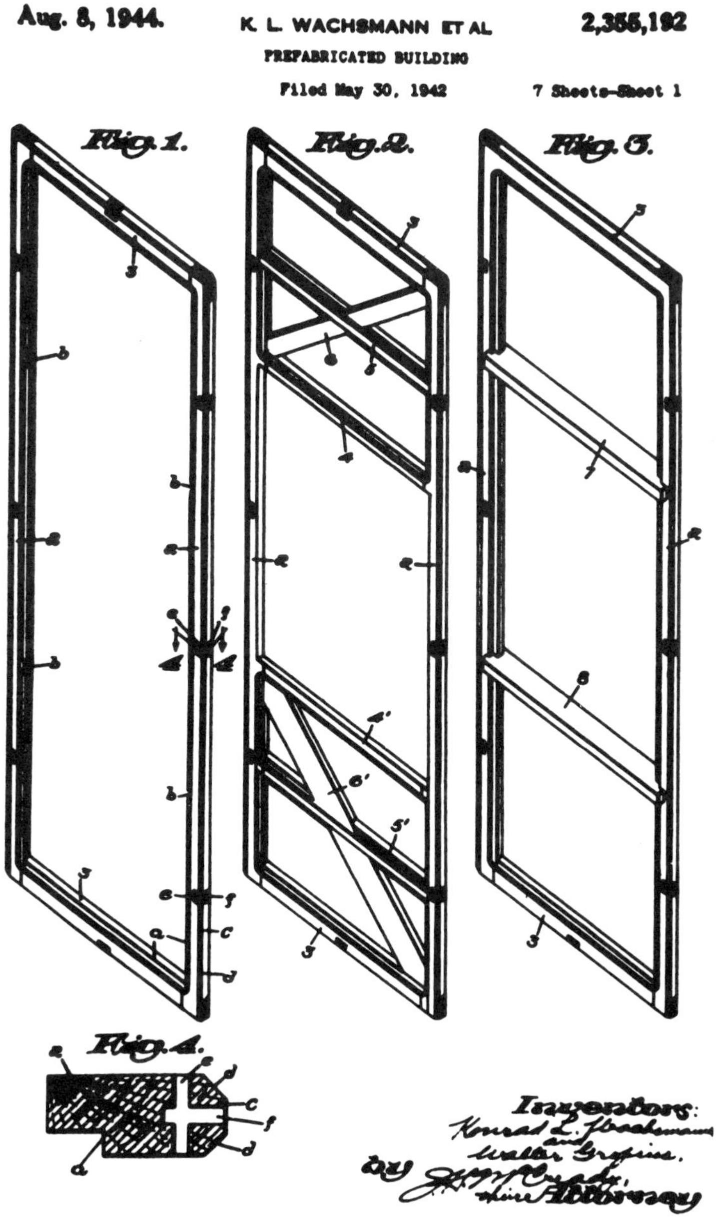

8.9
Wachsmann et al., prefabricated building (U.S. patent application 1942)

construction system, whose limited set of integrated components all stemmed from a single design source and ultimately would all have to be produced in one comprehensive factory.

On the other hand, as a design system in relation to the end product produced, it was conceived essentially in much more flexible terms. Here it could be described as open-ended. It did not postulate a standard design, nor did it even envisage a standard set of house designs. It was intended to generate a very wide range of design options, which could not, in fact should not, be predicated in advance. These design options were of course not infinite but limited by the parameters of the construction system to a family of designs that were all rectilinear, modular, panelized, and low-rise.

At this stage, whether considered as a construction system or a design system, it was essentially product oriented. The factory-made house is in itself a subsystem of a much wider system, which goes far beyond the physical object, and that is the housing process. The factory-made house is a product; industrialized housing is a process. In that process there are of necessity other subsystems, relating production to distribution, financing, legislation, transport, and land. The Packaged House, as a product to be manufactured, was not yet conceived in terms of this wider context.

As a tool for generating houses, and as a physical product, the Packaged House was a unique and advanced conception, and yet, like many other great inventions, its uniqueness lay in an original synthesis of known and well-established elements. Except for the four-way metal connector (correctly claimed in the patent application as "new"), all the other aspects of the system were, by 1942, quite well known in theory and had often been demonstrated in practice. Load-bearing wood-framed panels, independent of a structural frame, had been used previously by both Wachsmann and Gropius. Wachsmann had worked with modular load-bearing panels at Christoph and Unmack prior to 1929, and Gropius had employed them in the Hirsch copper houses,[15] based on the Förster and Krafft patent whose beveled edges were almost identical to the prototype design brought by Wachsmann from France. There were also several such systems used in the Growing House exhibition in Berlin in 1932, which had received wide publicity.[16] An interesting parallel in the United States was the Modulok system[17] of Arnold Southwell, an architect who was later to become involved in the Packaged House story. As far as the universality of the system was concerned, it is of interest to note an earlier proposal by architect E. Friberger for the Toreboda system, in which he used 3 m × 1 m panels which were standard except for their finishes for floors, roofs, and walls.[18]

Only the ingenious four-way metal connector of Wachsmann was entirely original. And yet, even here, it is not altogether without precedent. Some examples are particularly relevant. The infill panels of the Uninorm system of Constructions, Demontables Uninorm, of Paris, of 1938, were connected by a metal fastener. The "panels are fixed to each other by a special locking device with a key, two to each panel . . . an interesting locking device permitting rapid demountability."[19] As one of the principal uses of the system was for temporary shelters and barracks, one wonders if Wachsmann perhaps came across the method when he was interned in France in 1939 and acted as the camp's "director of building operations."[20] Other examples are even closer to home. The Hirsch system locked the wall panels together by bolting them to a small vertical steel channel section. Gropius sought to eliminate this steel element by housing the fastenings directly in the wooden panel frame. To this end he designed in 1931–32 a most ingenious spring-loaded metal fastener,[21] which was, however, not adopted by Hirsch. And then the Christoph and Unmack system, as we have seen, connected the panels by means of metal catches, four on each side, let into the framework of the panels, and had in fact been using metal fasteners (hooks, bolts, clamps) from the earliest days.

If attention is drawn to these precedents here, it is not to detract from the very real contribution made by the Packaged House, in both its advanced design of every separate element and the overall consistency of its general conception. Our purpose is rather to show that, inevitably, the Packaged House was a product of its times, the climax to the evolution of prefabrication in previous decades. As such, it grew naturally out of the rich experience that had been garnered by both Gropius and Wachsmann in their separate paths to mastery in the field.

The Contributions of Gropius and Wachsmann

There is no doubt, from the historical evidence, that the Packaged House was initially the brain child of Konrad Wachsmann. Gropius confirmed this with characteristic generosity, ascribing to Wachsmann a "decisive part in the scheme."[22] However, in saying this, he nevertheless went on to claim that, in the development of the Packaged House, he and Wachsmann had "pooled our experiences."[23] This is literally correct, as in the evenings they mulled together over the principles and the evolving details of the scheme. But it is perhaps also true in a more general, and much more significant, sense. The original Wachsmann proposal, the French scheme, which is the prototype of all subsequent mutations and developments, is the product not only of Wachsmann's ingenuity but of a whole decade of experience of prefabrication to which Gropius in Germany had given the prime theoretical direction and a great deal of practical impetus.

From this point on their individual contributions to the joint endeavor varied in kind and in quantity. Wachsmann's contribution to its technical development predominated, and most of the drawing-board decisions—those decisions of detail that are critical to architectural projects—were made by him. Gropius of course could contribute significantly here. He had a firm grasp of technical detail, and even though his experience in Germany had been predominantly in metals and lightweight construction, he had built up an extensive experience of building in wood since coming to America.[24] But because of the limited time at his disposal his role was more that of critic and sounding board: Wachsmann, working full time, was the essential innovator of detail. Gropius, on the other hand, provided the theoretical framework which gave architectural meaning and a human goal to the technical means: the conceptual framework of his philosophy of unity and variety, flexibility and growth, stability and change, standardization and individual choice. He also gave the Packaged House its first architectural form with a set of drawings showing possible house types generated by the system—a two-family house and a double-story row house with dwelling units of various sizes.[25]

Each in his own way responded to the challenges of technology and industrialization. Hence Wachsmann, at the Princeton Conference on Building for Modern Man:

The planning of man's physical environment has to be based on the best use of the available technique, which in turn is based on our knowledge of and our ability to control energy; in other words, on our economy and on our science. Only when it uses such means can a building, in

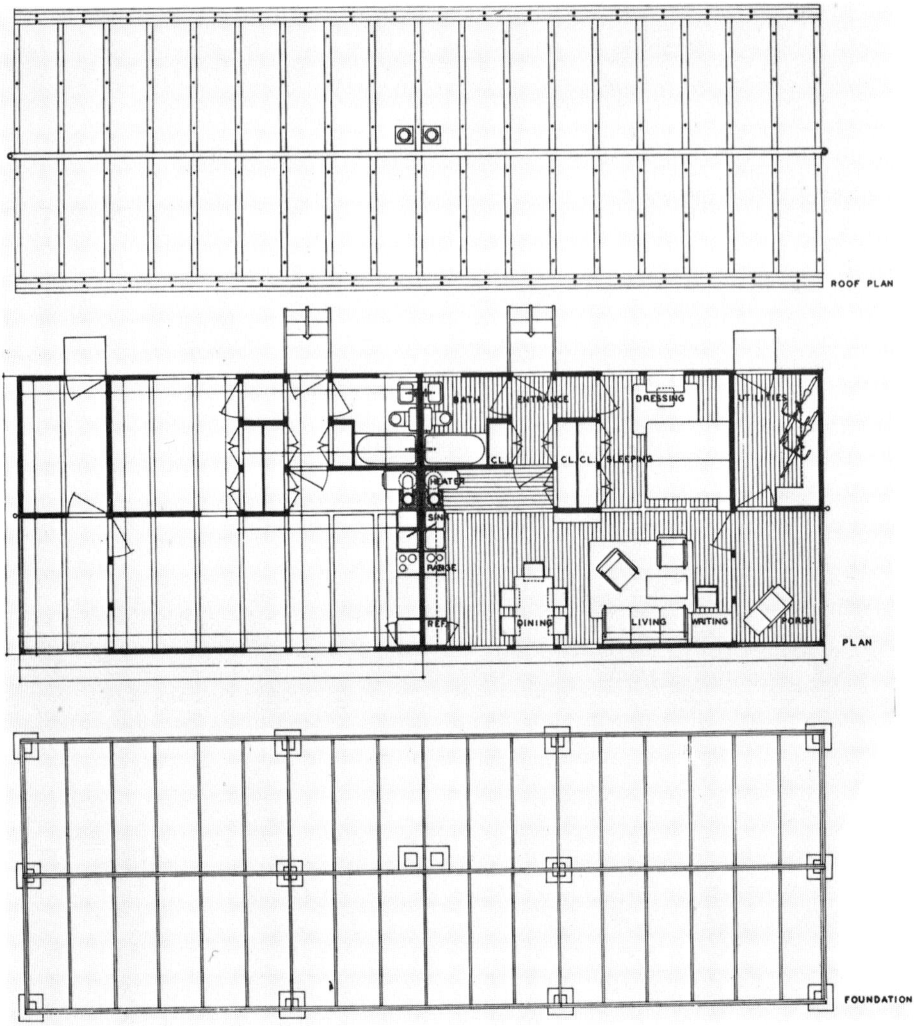

8.10
Gropius and Wachsmann, Packaged House, standard house type A, 1942

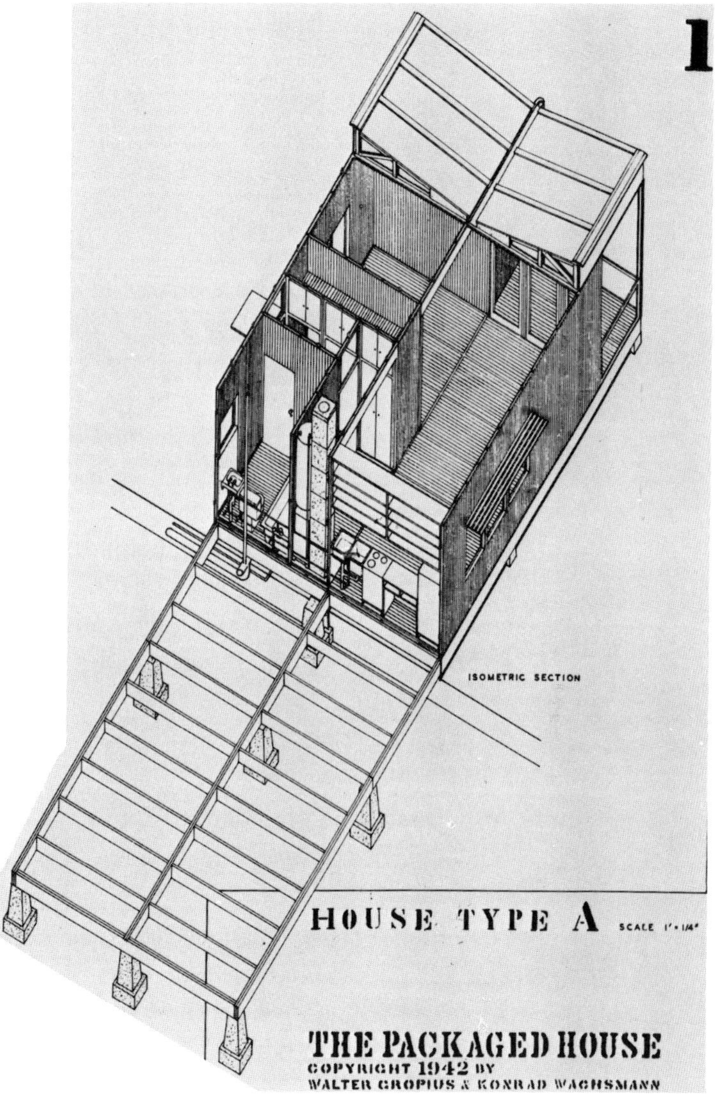
1
ISOMETRIC SECTION
HOUSE TYPE A
SCALE 1' = 1/4"
THE PACKAGED HOUSE
COPYRIGHT 1942 BY
WALTER GROPIUS & KONRAD WACHSMANN

any age, be called modern. Anybody who is able to improve *such methods, even in abstract terms, is indeed an artist.*[26]

Gropius, on the same occasion, had this to say:

Men will always rebel at attempts at overmechanization which are contrary to life. But industrialization will not stop at the threshold of building. We have no other choice but to accept the challenge of the machine in all fields of production until men finally adapt it fully to serve their biological needs.[27]

In other words, to Wachsmann the technological imperative was sheer poetry, by which man, in heroic terms, mastered the universe, whereas to Gropius it was an inescapable force, a means to be transmuted by man into serving his human goals. Gropius was in some sense a nineteenth-century figure, a humanist with an ambivalent attitude to the machine, fearing its dehumanizing potential but recognizing it, in Etienne Cabet's phrase, as "humanity's emancipator";[28] Wachsmann, of a later generation, was a man of the twentieth century, glorifying in technology and the science which underlay it as a source of light and poetry. Despite these fundamental differences they shared some important values:

An intense involvement in people and problems, a quality that Buckminster Fuller defined in them as love.

An unassailable optimism and a forward-looking faith, expressed in Wachsmann's phrase, "The future is everything."

A synthetic vision, always taking the comprehensive view and seeing potentialities for relationships where others saw only boundaries and incompatibilities.

In the fateful months between December 1941 and February 1942, within an ideology jointly conceived, Wachsmann produced the superbly drafted folio of drawings, and Gropius provided the logistic support: his home as drafting space and a base for operations; the capital, which provided for Wachsmann's work and livelihood and the cost of preparing drawings and models; access to legal guidance for making the patent application. Most important of all, Gropius operated within an incomparable network of connections which he had by now established through his formidable international reputation, his high standing at Harvard, and his inherent qualities as a human being of warmth and integrity. Such a network provided access for the Packaged House proposal to sources of influence in the press, the government, the academic world, and even the fringes of high finance.[29]

They entered the affair as equal partners and cooperated willingly, selflessly. But the stresses under which they operated were great. There were the macrostresses of the troubled world situation: it was not easy to be categorized, even if only in a technical sense, as enemy aliens in wartime America;[30] it was not easy, in the tranquillity of New England, to ponder the fate of their friends and family in Germany. And then there were the microstresses: the eternal worry about money, the tensions of an overlong stay as a houseguest, where hospitality, however generous, eventually becomes a burden on the receiver as well as the donor, and frictions caused by personality differences begin to arise.

Not only was Ise Gropius by now, much to her husband's concern, beginning to become restive with the situation, but by the spring of 1942, it was becoming apparent that the future of the Packaged House system lay not in the secluded world of Gropius' house in Lincoln, nor in the academic ambience of Cambridge, but in the hurly-burly financial world of New York. The hard decision was taken for Wachsmann to leave the drawing studio for New York City. Ise Gropius, whom he greatly admired but with whom, in recent weeks, he had increasingly crossed swords, bade him a hero's farewell. "Come back to Lincoln," she said, "either with your shield, or on it."[31] Wachsmann left Gropius, as he had arrived, practically penniless. He still spoke little English and faced considerable hardship in New York. But overcoming difficulties was a way of life with him; he was stimulated to be a free agent again and was optimistic that he would successfully meet all challenges. Great events had taken place in recent months, not only in the studio at Lincoln but also in the public arena, in the world of housing. On 24 February 1942 President Roosevelt "used his war powers to consolidate all Federal housing functions with a new National Housing Agency under a single administrator with full powers."[32] In the same month it was reported that a division of the FWA had allocated $153 million for demountable housing, in a vast program to house defense workers relocated through the decentralization of industry. The program, which aimed at producing 42,000 dwellings, provided a great oppportunity to the prefabrication industry.[33] Gropius' evidence to the Select Committee of Congress had possibly been a factor in this highly favorable development; Wachsmann would have to hurry, however, if the newly invented Packaged House was to be ready in time, to share in this promised windfall.

9

The General Panel Corporation

General Panel Corporation, New York

Wachsmann labored incessantly and uncomfortably in the unfamiliar heat of a New York summer, to complete his work: to prepare all the details and to make a model. He was now almost completely out of funds, and becoming increasingly desperate. Despite the intense pressure to finalize the project and find a backer, he was, at the same time, trying to revise and improve the system. The metal connector, which had undergone several transformations from the original Y-shaped element to the patented wedge connector, was once more under Wachsmann's scrutiny, and a new concept began slowly to emerge.[1] He knew he should be attending to more urgent matters, but the fire of perfectionism which burned in him was unquenchable, being extinguished neither by financial care nor by pressing immediate concerns.

At the very nadir of his fortunes, in September 1942, in a chance storybook encounter pregnant with improbability, he made the acquaintance of Jack Marqusee, of the Wall Street firm of investment bankers, Charles Allen and Co., and succeeded in interesting him in the potentialities of the Packaged House system. Through Marqusee's intervention he was able to put his proposals to Charles Allen and Co., and in a tour de force of persuasiveness, his enthusiasm and faith in the system overcoming the barriers of an interpreter,[2] he convinced the chairman and the board to back his and Gropius' scheme. Within days the fate of Packaged House system underwent a radical transformation. A corporation "to manufacture, purchase, import and otherwise acquire, export, sell or otherwise dispose of, design, handle, traffic and deal in, erect, construct and assemble prefabricated houses, panels, partitions and the like" was registered in New York on 12 September 1942, under the title of the General Panel Corporation, with Roy Plaut (of Plaut and Schweitzer), Jack Marqusee, and Halsey D. Josephson as first directors and stockholders.[3] Wachsmann found himself on the payroll, with a handsome salary, and the possessor of a check for $10,000[4] for the construction of a demonstration house, which, it was hoped, would be completed within three months.

Five months later, on 23 February 1943, the house was shown to an invited group at the Somerville, Massachusetts, premises of the U.S. Plywood Corporation.[5] Gropius set up the demonstration and invited a large and influential gathering of government officials from Washington, representing the FPHA (whose Technical Department had already approved the system) and the military, together with interested architects, engineers, and contractors. This gathering was duly impressed when the demonstration house was erected, and then taken down again, all in one day. This house, which had been handcrafted in Boston, was planned in accordance with the space standards and requirements of the NHA's type TDU-1, a temporary dwelling unit for which Wachsmann had been preparing drawings in New York since January 1943. Also completed at this time were the drawings for a prefabricated barracks building, using General Panel standard components.[6] Gropius and Wachsmann were thrilled at the progress being made, and there was talk of going immediately into production. Application was made by the voting shareholders, now eight in number, to increase the maximum number of the directors of the corporation and to increase the number of shares issued.[7]

Now came the time to bring the work of General Panel to the attention of the public, and a massive public relations campaign was undertaken, probably by Walter Gropius who had all the right connections. In April there were long and copiously illustrated articles in the architectural press: by Willo von Moltke (a former student of Gropius' at Harvard) in the *Architectural Record,*[8] and by Herman Herrey in *Pencil Points.*[9] This was followed in June by a review article on prefabrication in the *Architectural Forum,*[10] which made generous reference to the General Panel house. This extensive coverage spread to the lay papers and journals, *Christian Science Monitor, Colliers, Business Week, Saturday Evening Post,*[11] in articles and advertisements.

After the frenzy of 1942 came a time of consolidation, and by mid-1943 activity had slowed down considerably—or perhaps more accurately, the output of activity had become less visible.[12] Gropius, who at no time drew a salary from General Panel, although both he and Wachsmann were vice-presidents, gradually became more detached from the affairs of the corporation. He visited New York from time to time, looked over the shoulders of the draftsmen with interest, and made informed and useful critical comment.[13] But in the day to day development, he now stood on the sidelines, as it were. He still saw a useful role for himself in exploring the design potentials of the system, and at the end of the year his students undertook a studio project at Harvard to investigate the General Panel system's inherent qualities of flexibility and variability.[14] In this concern with the architectural exploitation of the system, Gropius was

9.1
General Panel, New York, demonstration house, Somerville, Massachusetts, 1943

9.2
General Panel, New York, sectionalized house, type TDU-1, 1943

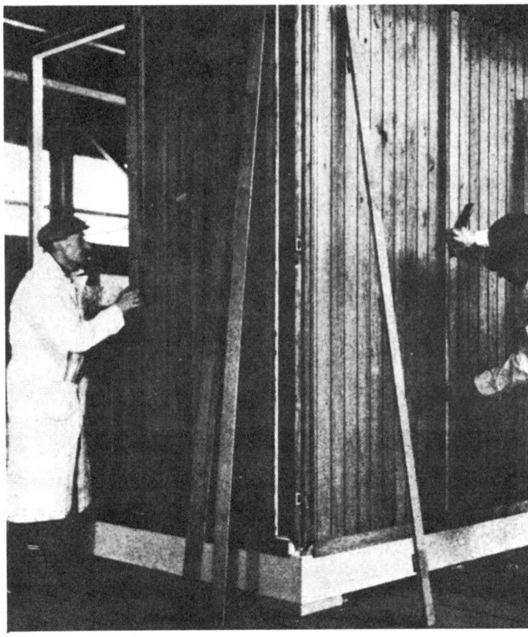

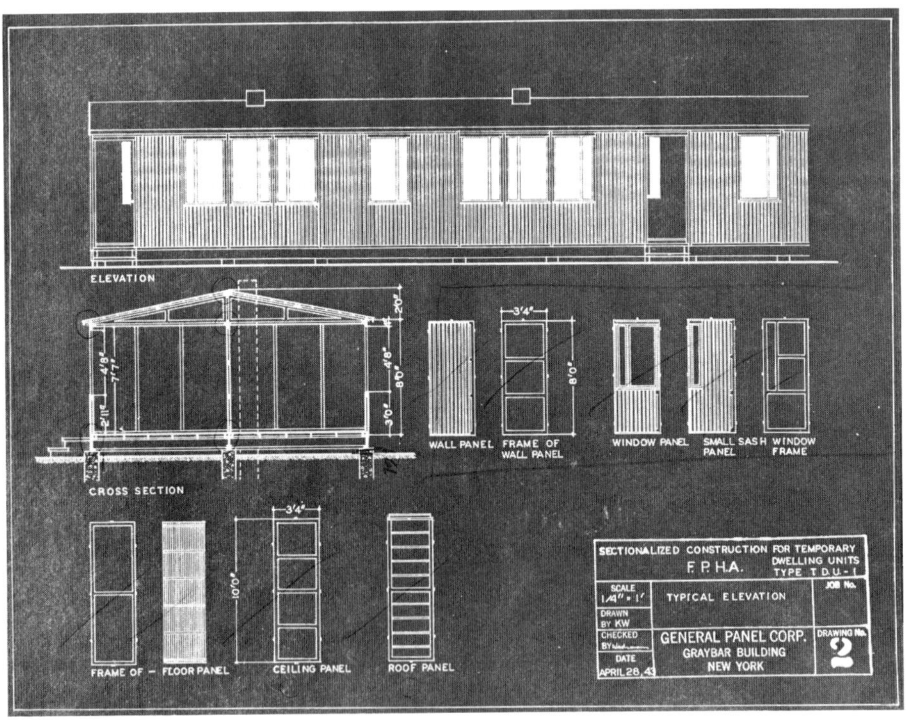

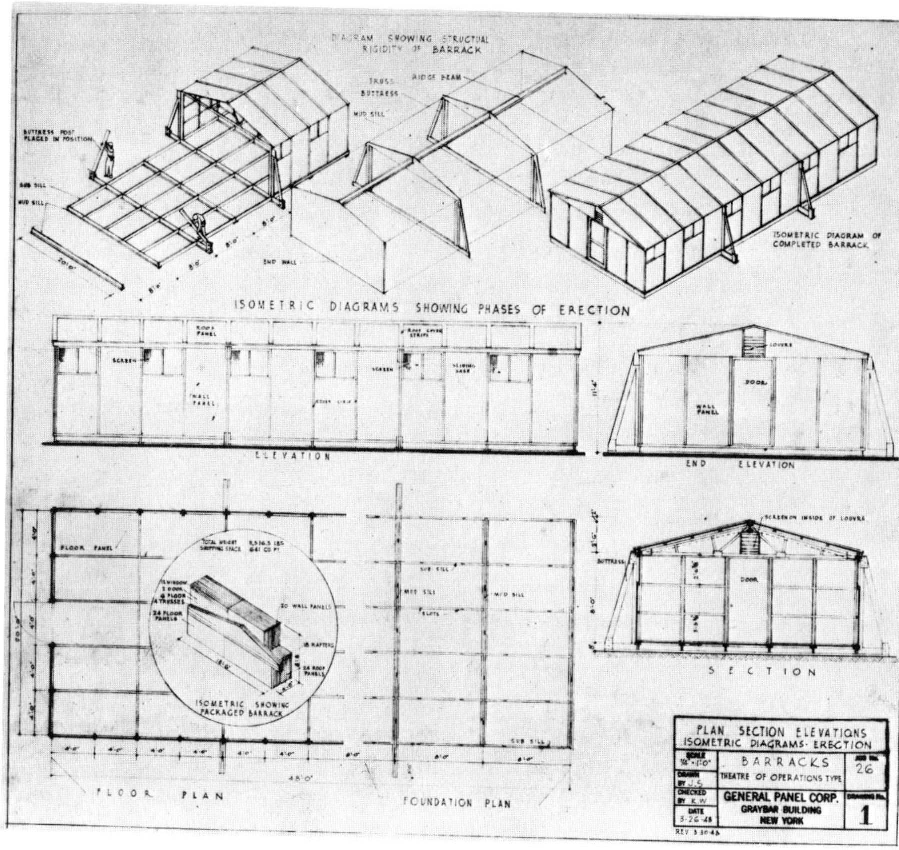

9.3
General Panel, New York, sectionalized barracks, 1943

dealing with the challenge he had been postulating since 1910: How much variety can be generated by a system of standardized parts? This was the thrust of his students' work and of a later set of drawings prepared under his own direction, labeled the "Flexible House."[15] It was upon this design theme of the "expansible house" that the *Architectural Record* commented: "For the many new families who will be impatient to set up their own new homes after the war, Gropius has drawn this as one suggestion for a minimum-cube, expansible house. It could be built in quick time via the prefabrication system devised by Wachsmann and Gropius for the General Panel Corp. . . ."[16] In these projects Gropius continued to stress that linkage between prefabrication and the concept of flexibility, which had all along been one of his main theoretical postulates. Wachsmann, on the other hand, concentrated at this time on the technical development of the system, and the interpretation of standard building types in terms of its construction details and modular discipline.[17] But he too, like Gropius, was not now giving his undivided attention to the Packaged House project. If in 1943 efforts on behalf of the Packaged House system seemed to diminish, then in 1944 progress toward the main goal—the actual manufacture of the house on a commercial basis—was even less. For this relative stagnation there were two principal causes: a lack of adequate capital for development and the deflection of Wachsmann's interest into two alternate projects. To the question of finance we will return later, but we must first examine Wachsmann's alternative interests at this time. The first was a project related to the General Panel Corporation but not to the Packaged House project. During 1943, while the Packaged House marked time, as it were, Wachsmann had been developing a modular panel office partition system, based on an ingenious J-shaped interlocking joint system. In December 1943, an application was made by Wachsmann for a patent for a "sectional wall structure system,"[18] the rights to be assigned to the General Panel Corporation of New York. It must be noted here that Wachsmann and Gropius had entered into contracts with the General Panel Corporation to assign to the corporation all future patents on any related design, until certain specified dates.[19] To the development of this partition system Wachsmann now devoted much time and energy; it was a new problem, and as such fascinated him. His efforts resulted in a most impressive folio of 150 drawings, imaginatively conceived, immaculately detailed, precisely drawn, the preparation of which took up to the end of April 1944.[20]

In that month the Atlas Aircraft Products Corporation entered into two short-term contracts with General Panel, the one related to the development of that corporation's inventions, the other pertaining to the management of General Panel.[21] Both Charles Wohlstetter, president of Atlas, and his brother Albert Wohlstetter joined General Panel's board of directors. This new factor in General Panel came about through a connection between Wachsmann and Atlas Aircraft Products in an altogether different venture. Atlas had undertaken the development of that other great invention of Wachsmann's, the Mobilar tubular steel prefabricated hangar system.[22] It will be recalled that the origins of Mobilar were prewar and, like those of the prototype of the Packaged House, lay in Wachsmann's troubled but productive stay in France.[23] But unlike the Packaged House, Mobilar was Wachsmann's personal interest and did not fall into the scope of the General Panel Corporation. It was to the development of this system that he gave the greater part of his attention in the latter part of 1944 and the early months of 1945, completing a monumental folio of general and detailed drawings for the Atlas Aircraft Products Corporation by March 1945.[24]

9.4
Harvard Graduate School of Design, prefabs using General Panel components, 1943

9.5
Walter Gropius, expansible house using General Panel components, 1944

9.6
Wachsmann and Gropius, General Panel office partition system, detail of jointing, 1943–44

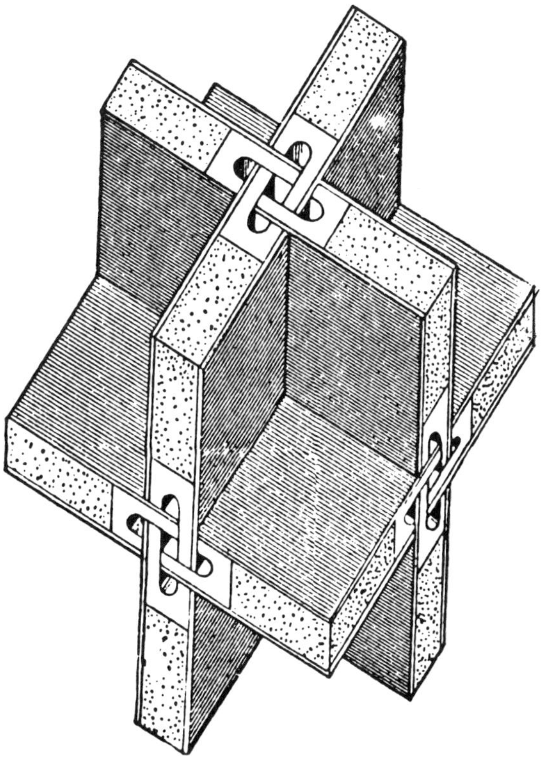

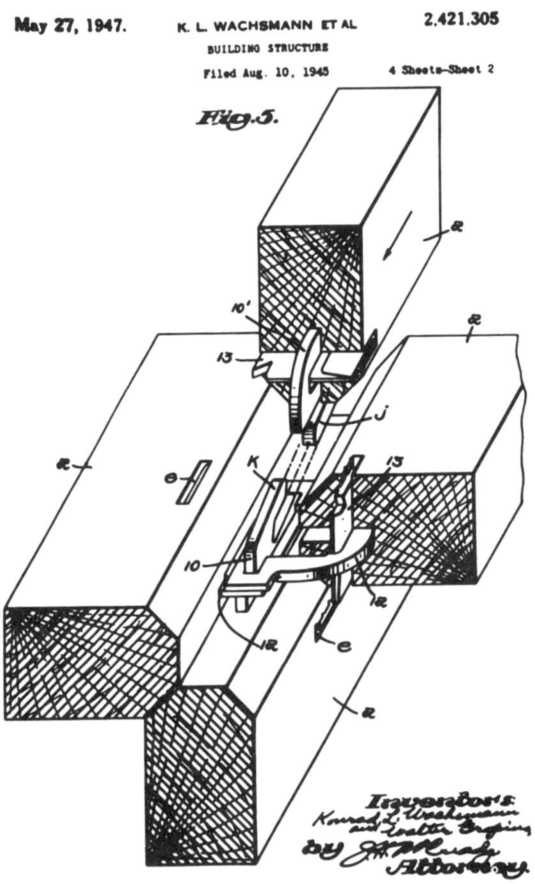

9.7
Konrad Wachsmann Mobilar hangar system, 1939–45

9.8
Wachsmann et al., building structure (U.S. Patent application), 1945

The only creative effort directly linked to General Panel in the latter part of 1944 was Wachsmann's refinement and perfection of his new metal connector. It will be recalled that in 1942, even as he and Gropius were filing a patent application for his wedge connector, Wachsmann had been preoccupied with a new system of jointing. Now in 1944, as the original patent was approved,[25] he was busy finalizing the design of its replacement. From August 1944 to January 1945, in the drawing offices of Atlas Aircraft Products and of General Panel in New York, the working drawings of the new connector were finally prepared for the machine-tool makers.[26] Wachsmann was drawn, almost compulsively, to this problem, and this metal coupling is by far the most elegant and ingenious of the whole series. The advantage of this revised method is stated clearly in the patent application of August 1945:[27] "This invention aims to improve the wedge connector . . . in such a manner that all these connected elements can be installed in the panels, or other building units, at the factory, thus simplifying the assembling operation at the site of the building and reducing the labor of assembly." Moreover the old system necessitated a cover strip, matched to the siding, to mask the connections; the new one did away with this, making it possible to use a virtually jointless flush wall, employing plywood.[28]

From War to Peace: The Search for Backing

The war in Europe ended in May 1945, in Asia in September. The Packaged House, conceived as a response to the crisis of war at the time of Pearl Harbor, had by the war's end still not gone into production. Amazingly, despite the enthusiasm with which it was initiated and pursued, despite the ingenuity of its design, despite the energy of Wachsmann and the reputation of Gropius, despite professional acclaim and governmental approval, despite its initial Wall Street backing, despite its subsequent linkage with experienced industrial management, despite all this it had missed the entire phase of wartime demand. America had engaged all its efforts in the war, peace had now come, and the best a prospectus of the corporation could say, at the beginning of 1946, was: "No houses have as yet been manufactured and sold although experimental houses have been manufactured."[29] Four years of intensive effort, and not one house built for sale—and this during wartime, with its incredible demand for instant housing, when the location of wartime industry involved a population shift of some 8 to 10 million workers,[30] this at a time when the climate of urgency was highly supportive of creative initiative, particularly in the industrial field. This failure of General Panel to achieve production must also be seen against the success of the prefabrication industry as a whole in these favorable times, when, with over 70 firms active (many with a production rate of 1,000 units or more per month), a total of over 200,000 prefabricated units had been manufactured during the war, more than half financed by public funds under the Lanham Act.[31] By aiming at technical perfection, a sound financial structure, an assured distribution system, proper production facilities,[32] and public approval, the directors of the General Panel Corporation, from 1942 to 1946, had no doubt sought a sound basis for the future but at the cost of present advantage. Their energies moreover were dissipated in the pursuit of other schemes, the partition project and the Mobilar hangar, and they were thus deflected from the main target. The brutal fact remains: in terms of exploiting the unique opportunities of the wartime market, the General Panel Corporation had, as it were, missed the bus.

In September 1945 as the war officially ended, a contract was entered into whereby the American Wire Fabrics Corporation, a subsidiary of Colorado Fuel and Iron Corporation, acquired "the exclusive right to manufacture and sell throughout the United States and its territories, dependencies and possessions, prefabricated buildings, panels, connectors and office partitions," made under the patented systems of General Panel. For this privilege American Wire contracted to pay a considerable sum in royalties during 1946 and 1947, and in addition guaranteed by 1947 "to manufacture and sell a minimum of 1,000 prefabricated houses a year."[33] Colorado Fuel and Iron had previously had interests in prefabrication, having sponsored a steel-frame system in the late 1920s.[34] Now they were becoming involved, through their subsidiary, in an entirely wooden system, but it was probably the metal connector that attracted them. And they envisaged additional income from the manufacture of this critical small item which would presumably be needed in large numbers.[35]

This infusion of new hope and of new capital came at a critical time. One of the main reasons for the slow progress of the project, after its initial flying start, was indeed the perpetual shortage of funds. The development of the system was proving to be a most costly business,[36] for which the initial investment of the Allen company and the private stockholders was entirely inadequate. The experimental house, demonstrated first in Somerville, then erected in New York, had cost $67,643; the experimental partitions, $50,852; the patent rights, over $15,000.[37] There had been much frantic searching, during 1944 and 1945, for new sources of capital. In this search, Jose (Pepi) Weissberger, a businessman with international connections and a close friend of Walter Gropius', had been indefatigable, trying to establish new markets and to attract new sources of capital: the Banco de la Propiedad in Mexico, Lord Donegall in London, Goodyear and the George A. Fuller Co. in the United States.[38] Now with the new agreement there seemed at last to be a firm financial basis to the whole venture, an arrangement that had the additional advantage of lifting the functions of manufacture and distribution from the already overburdened shoulders of Wachsmann, who was by this time the president of General Panel in New York.[39] Design and documentation, carried out with a compulsive attention to detail, fully occupied his time. He had established a relatively large drawing office in New York, under the direction of the Assistant Technical Director Curtis Fremond, and with Rudy Wolf as chief draftsman.[40] A set of three volumes of General Panel standard details had been prepared: connector and connection; wall, partition; windows, doors.[41] Booklets were prepared for possible export to Sweden and Canada.[42]

What was now needed, if this design effort was to be translated into action, was a massive further public investment. At the beginning of 1946 the corporation applied to increase its shares to a par value of $395,000, of which $300,000 was to be a newly issued preferred stock and $95,000 common stock (with voting rights); of the latter figure $80,000 represented the shares issued up till then to all shareholders.[43] This stock was offered to the public in an impressive prospectus; a fivefold increase in capital was thus sought, whose purpose was "to assure the Corporation of sufficient capital to enable it to conduct further research and to develop new inventions and patents in the prefabricated housing field and to develop markets for its products both in this country and abroad."[44]

With these new financial arrangements General Panel, having missed the war, was apparently now ready for the challenges of the postwar era. Indeed, it seemed that they, together with other prefabricators, were about to be given a second chance. The war ended, the soldiers returned, workers in redundant war industries relocated themselves: suddenly, once again, housing became a burning public issue, with demand high and supply short. In the face of what was evidently once more an emergency housing situation of crisis dimensions, the new president of the United States, Harry Truman, on 26 January 1946, "issued an executive order establishing the office of Housing Expediter charged with the task of preparing plans and programs and recommending legislation for the provision of housing for veterans. He named to this post Wilson W. Wyatt . . . ."[45] Wyatt moved to confront the housing crisis with great speed. By February he was able to put forward his Veteran's Emergency Housing Program, which not only strengthened the policy on price controls and rentals but actively stimulated the production of new materials, gave priorities to veterans' housing, and postponed all deferrable and nonessential buildings. As part of this program, "he proposed a large expansion in factory fabrication of houses through allocations of surplus war plants and materials and through guaranteeing the market for the product." His target, that originally proclaimed by the Senate Subcommittee on Housing and Urban Development in August 1945, was an initial production of 1.2 million dwellings for 1946, with increases the following year.[46]

A spirit of optimism once again pervaded General Panel. Two test houses were erected in Queens, near the La Guardia airport.[47] Gropius joined Wachsmann to inspect the procedures and was greatly pleased with the result. A careful photographic record was made of all stages of erection and of the final products, and an impressive color film made of the whole procedure. Houses for the system were designed by Wachsmann, Gropius, Richard Neutra, and others,[48] and the architectural press was advised of plans to go into immediate production, with "a starting volume of at least 3,000 houses a year," and of intentions "to build factories on the West Coast, in Colorado and in New York."[49] After four years of preparation, the center of gravity would now shift westward from New York to California, and the phase of active production, or so it seemed, would now begin. In order to achieve this, however, a new corporate structure was needed.

In mid-1945, other negotiations were taking place, which were to be of considerable importance to General Panel. Gropius, who was not always kept in the picture, and resented it, wrote somewhat testily to Wachsmann: "Meanwhile I am pretty much out of touch as to what has happened to the Corporation, whether a deal with Mr. Swenson is in the making or what trend our recent discussions have taken."[50] George E. Swenson, to whom Gropius was here referring, had, in 1936, invented a patent sandwich panel comprising outer skins of asbestos sheeting and a core of two or three fiberboard sheets, made by the Celotex Corporation. Persuaded that there was a future in this new building component (Cemesto, as it was called), Celotex decided to back it; after an uncertain start that future looked brighter when the John B. Pierce Foundation used it as a substitute for plywood in its experimental horizontal construction system for prefabricated houses. This system of construction, we may note in passing, had been of some interest to Walter Gropius at the time the Packaged House was being evolved.[51] It had passed from an experimental system to reality in 1941 when a very large housing scheme designed by Skidmore, Owings and Merrill for the Glenn L. Martin Corp. was constructed in Baltimore, utilizing the Cemesto boards in the Pierce system.[52] By 1943, responding to the challenge and the opportunity of defense housing, Celotex was producing, in two factories, sufficient panels for 1,500 houses a month.[53]

9.9
Walter Gropius and Konrad Wachsmann, inspecting test house, General Panel, Queens, N.Y., 1946

9.10
Wachsmann and Gropius, General Panel test house, Queens, N.Y., 1946. *Top*: under erection; *bottom*: completed house.

9.11
J. B. Pierce Foundation, with Skidmore, Owings, and Merrill, housing for Glenn L. Martin Corp., 1941

9.12
Gidoni, Bromberg, Gropius, Neutra, designs for houses using General Panel components. 1947

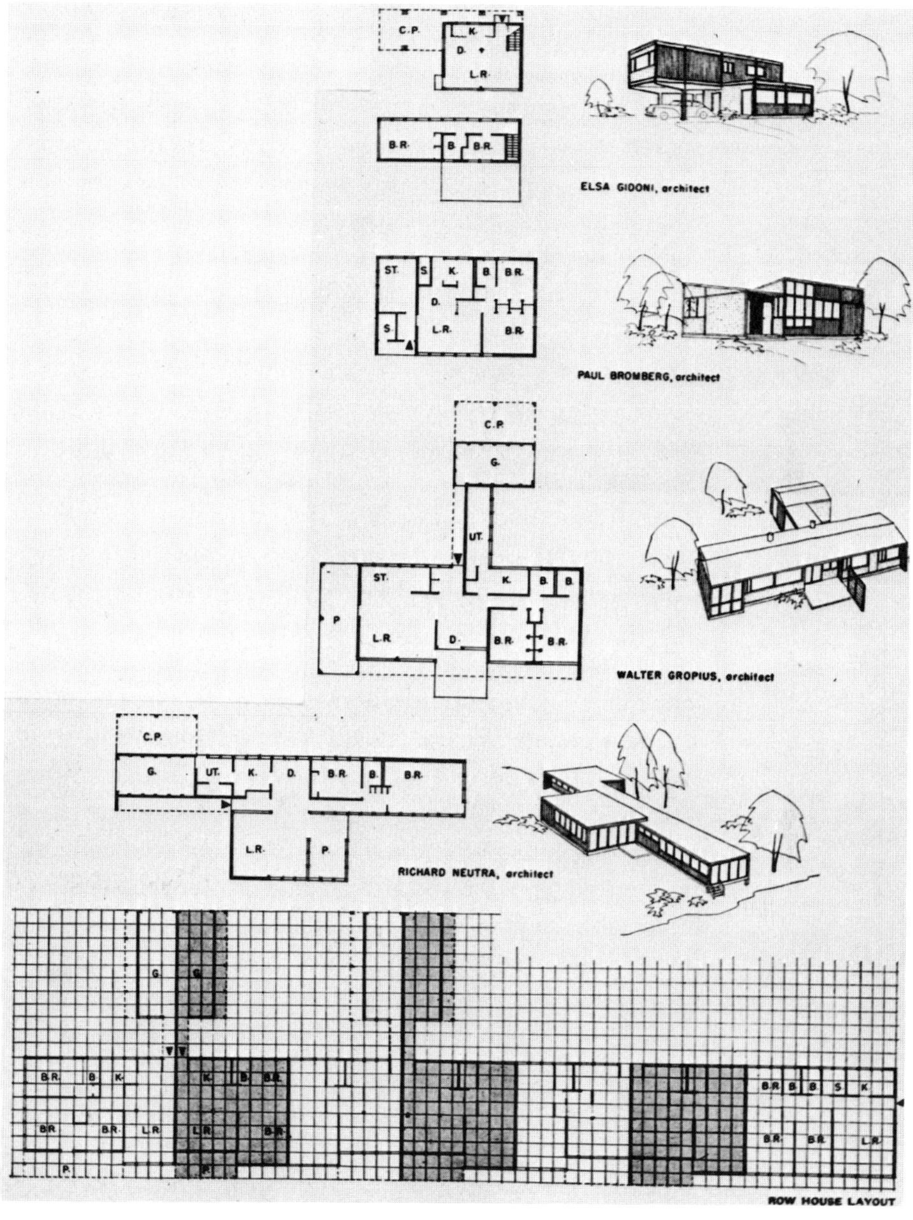

At the time of the negotiations with General Panel, Swenson was assistant to Bror Dahlberg, the President of Celotex. Dahlberg, an industrialist of international stature, was a formidable advocate of solving the postwar housing problem through the mass-marketing of prefabricated houses.[54] Although Celotex had a direct financial interest in the widespread use of new methods and new materials, Dahlberg's faith in prefabrication went beyond the immediate vested interests of his corporation. His son, Carl B. Dahlberg, apparently shared these interests, and was connected with two associated companies, the Drycemble Corporation of Houston, Texas, and Modulok Incorporated of Baltimore, Maryland.[55]

The Celotex Corporation had apparently long had an interest in the Packaged House. In 1943 it had illustrated the design by Gropius of a prefabricated house using the General Panel system in one of its "Celotex Miracle Home" advertisements and had reported it in its own house journal, *Celotex News*.[56] Now in 1945, Swenson of Celotex and the Board of General Panel were quietly exploring the possiblity of a business relationship.[57] Celotex had a major short-term interest: to move General Panel into the productive phase and to have the main hand in the distribution of its products. A longer-term objective was doubtless to find a place for Celotex-produced materials in the General Panel system.[58] These objectives were not incompatible with the interests of General Panel, and by the middle of 1946, after a year of discussions, some important tangible results emerged. Largely on the initiative of the Dahlbergs, who were probably the major shareholders, a new corporation was set up: the General Panel Corporation of California.[59] This was registered as a "Delaware corporation" in June 1946, with Carl B. Dahlberg, son of Bror Dahlberg, as president, and Albert Wohlstetter, brother of Charles Wohlstetter of Atlas Aircraft Products, as vice-president in charge of production. Other officers included the treasurer, Nathan H. Wendell, who was vice-president of the National Association of Housing Manufacturers (one of two organizations representing the interests of prefab manufacturers), and the technical director, Schuyler Southwell, sometimes referred to as chief architect. The Southwells, Arnold and Schuyler, had been associated with the Drycemble and Modulok companies, which were no longer in production. Wachsmann's position in the new corporation at this early stage is not clear, and the evidence is contradictory; he probably only became a director, and vice-president, in 1948.[60] Gropius had no official role in General Panel of California, although he remained a vice-president of General Panel, New York. The relationship between the two corporations is of critical importance to the ensuing history of General Panel. The New York firm, with Wachsmann at its head, was a design-and-development-oriented engineering corporation, which owned the Packaged House and General Panel partition system patents. The newly established California corporation was a manufacturing firm, which was to make the Packaged House under license, paying the New York firm a fee of 2 1/2 percent of net sales. Gropius and Wachsmann, the designers of the system, were to receive royalties from the New York corporation. Celotex's role, as defined in a contract being negotiated with General Panel of California, was to market the products of that company in the western states. Although the parties to these arrangements were separate and independent entities, there was yet a degree of functional and administrative overlapping. The New York company owned a 10 percent share in the California company; at various times directors (Wachsmann, Wohlstetter) served on both boards of directors; George Swenson, maintaining liaison with Celotex, eventually joined General Panel's New York board.

In the development field, this overlapping created a situation of potential confusion and friction. New York was to provide design and research services to California, for a fee, but California intended setting up its own technical research office under Southwell, with about four architects and engineers and ten assistants. One may be reasonably sure that Wachsmann was not happy with this division of authority: he certainly had grave misgivings about the direction the negotiations had taken.[61] Walter Gropius too had reasons for disquiet. He had long advocated—to AEG in 1910, to Hirsch Copper in 1931—that the manufacturing of prefabricated houses be separated from their marketing and finance, but he had always seen the closest link between design and manufacture. The new corporate structure of General Panel was not in accordance with this model. On the contrary, capital investment, manufacture, and sales were closely linked in the California venture, where the Celotex connection had a strong guiding hand; the designers in New York, led by Wachsmann, were in danger of being isolated, perhaps bypassed.

Climax in California: The Rush to Production

In accordance with Wyatt's policy of allocating surplus war plants to the manufacturers of houses, General Panel of California acquired from the War Assets Administration the former Lockheed Aircraft Corporation's engine factory at Burbank, California. This property, about 234,000 sq ft in extent, was purchased for $769,306, of which only 20 percent had to be put down, the remainder being covered by a mortgage from the War Assets Administration.[62] Even this limited downpayment was enough to deplete severely General Panel Corporation's initial capital of $400,000, and additional funds were sought. By the end of 1946 a construction loan was obtained from the Reconstruction Finance Corporation, General Panel being one of the only three manufacturers of prefabricated houses to receive this aid.[63] At the same time a guaranteed market contract from the government was sought to cover an initial production of 8,500 houses. The process of eliciting such a guarantee was long and protracted. Charles Wohlstetter had suggested it to Wachsmann, even before the new corporation was formally registered.[64] In September, because of the delays in finalizing the formal guarantee, Dahlberg requested from the NHA a letter of intent to enter into a guaranteed market contract, covering 8,500 General Panel Model B-16 Units (low-cost four-roomed houses) at a RFC purchase price of $2,612.60 each.[65] His proposed schedule, at that time, was for 1,000 houses in the first quarter of 1947, followed by 2,500 houses each subsequent quarter.[66] In December 1946 this letter of intent was received, after the NHA's housing expediter had certified "that the house as now designed with the minor changes made to comply with the suggestions of this office and the FHA constitutes a structure which is sound, durable and livable."[67]

Government support in the form of these loans and guarantees had been achieved by General Panel just in time. In the congressional elections of November 1946, the Republicans gained control of both houses of Congress. Truman, sensing the conservative mood of the country, began to trim his more liberal measures. When Wyatt put forward his new housing program, it was substantially rejected by the administration, and he had little alternative but to tender his resignation. Then, on 11 January 1947, Truman terminated the short-lived Veterans Emergency Housing Program.[68] Although the RFC loan program and the NHA market guarantees did not immediately come under the axe, it seemed likely to knowledgable observers that "these special helps to prefabers will also soon follow priorities and price controls into the federal wastebasket."[69]

At this critical point in national housing policies, General Panel of California was in the process of equipping its Burbank factory with the most sophisticated, complex, and expensive production plant. Together with some remodeling of the Lockheed factory, design of the plant layout had been going on intensively throughout the second half of 1946.[70] Most of the work was conceived, designed, and detailed by Wachsmann and his New York office,[71] although there was some participation by Southwell's department in Burbank, particularly in the correlation of physical layout with staff organization. "It is difficult to depict the tremendously detailed forethought which had obviously been put into [the] setting up of this organization . . . this has obviously taken a tremendous amount of research experience and highly intelligent thinking," reported an observer undertaking a survey of the prefabrication industry on behalf of the Bemis Foundation. He had made a field trip to the factory, where he had been shown, by Southwell, the highly detailed staff and plant organization charts, the plant layouts showing not only machinery but the exact position of each worker at his production station, and the time studies and estimates of quantities of materials.[72]

This was in February 1947, and though there had been reports the previous month that the installation of the production line had commenced,[73] there was obviously still much to be done. "All equipment is promised for sure," the man from the Bemis Foundation was assured, and General Panel were "expecting to be in production by the time predicted," that is, by 15 April 1947 when it was planned to commence with a production of 5 houses a day for the first 30 days, then 10 per day, stepping up to 40 per day by August.[74] By the April deadline, however, it was apparent that production was still at least a couple of months away. Those machines already ordered included key items whose delivery dates had not been met, and urgent discussions were being held to try and get delivery by mid-June, even at the expense of prohibitive overtime charges.[75] General Panel, in April and May, was still working on its own drawings of machinery whose manufacture had thus not even started. Wachsmann, in fact, reveled in the designing of the new plant and its production machinery. The problems of the Packaged House as originally conceived were by now long solved, and his creative spirit restless, he cast about for new challenges to his ingenuity. He was moreover no longer even sure about the validity of his original design as the best possible and, to his business partners' horror, would gladly have scratched the entire scheme and started all over.[76] By mid-June the supply and installation of the main press had not yet been finalized, and the impressive plant layout plan, "far beyond anything one has ever seen in this field," was still incomplete for lack of final details.[77]

Inevitably, production was once more deferred by a further month, with three alternative programs of production;[78] but for even the most modest of these options, a pilot program of five houses a day, starting 15 July, problems with the supply of materials were envisaged. Eventually, in July 1947, more than five years after the Packaged House was designed, but only a year after the establishment of the California company, production "on a limited basis"[79] commenced in a plant fully equipped to turn out 10,000 houses in a year, working on a single shift. The tale of successive delays detailed above can only but emphasize how considerable was this achievement. Within a year, in a burst of creative energy to which all the team contributed, but in which the driving inspiration was undoubtedly Wachsmann's, the layout of the productive process at Burbank was conceived in principle and designed in detail, and a highly sophisticated, controversial, in many respects innovative automated factory, with all its complex machines, was designed, negotiated for, manufactured, and installed to the point of the commencement of production. The creativity of this year paralleled that of the first year of the project, when the Packaged House was designed, perfected, and voluminously documented. But between the initial outpouring of design, and the final rush to production, there remains the inexplicable lacuna of two or three years, the lost years of New York when the General Panel project floundered without real progress and without clear direction.

General Panel, in July 1947, was at last set to go ahead, with the most splendid of technical installations, to produce the industrialized house. The planned production system envisaged a highly automated straight-line manufacture of standardized parts. In Wachsmann's design concept of these parts, the final system of manufacture was always adumbrated: product and production system were to be an organic whole, totally integrated. "The standardizing of our basic product makes them highly adapted to modern machine needs. We have equipped our plant with specially designed machinery of a high-speed, multiple-operation character; radio frequency installations which cure glue bonds in a matter of seconds; and advanced quality control devices which assure steady and reliable quality in the product of every work station. These processes are repetitive and highly adapted to straight-line production, since the universal joint means that all panels, whether for floors, walls, partitions, ceilings or roofs, are structurally similar from the standpoint of production."[80]

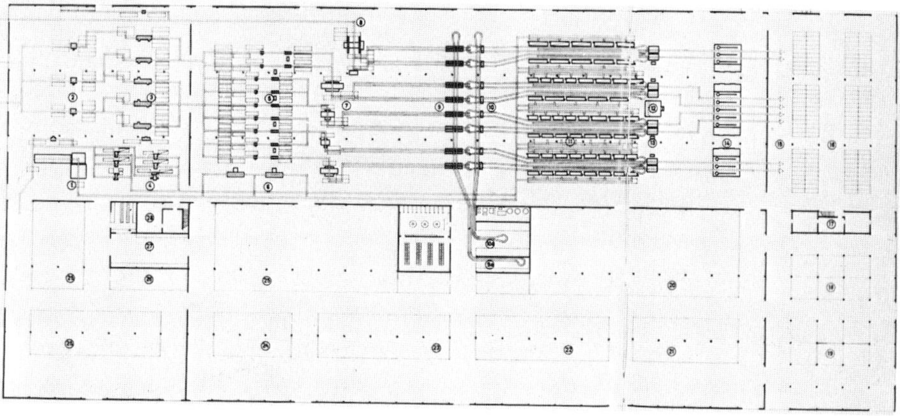

9.13
General Panel, California, layout plan of factory, Burbank, 1947

9.14
General Panel, California, factory, Burbank, milling area, 1947

From the mills,[81] the redwood timber, with moisture content carefully controlled, was carried by conveyors to a series of self-fed, belt- or hopper-fed machines: first the high-speed moulders, then the double-tenoning, coping, and multiple-mortising machines. All the components, the panel frames, the structural elements, the filler strips, were then taken to have their cadmium-plated die-cast steel connectors inserted. Frames went through glue spreaders and were then placed in the steel jigs; insulation where required was inserted; Douglas fir plywood sheets arrived by conveyor, to be tack-welded to the frames using specially designed high-frequency gluing equipment, the entire stressed skin panel then being bonded in an electronic press.[82] Finally, every element was sprayed with a pigmented prime and second coat, to await the final coat of paint on the site.[83] Stockpiles of panels, structural elements, and filler strips were stacked, sometimes in "house lots," ready to be loaded on to trucks and "shipped from the factory complete with insulation, sash, doors, glazing, hardware, trap-doors, vents, special openings, concealed wiring, connectors"[84] ready for final erection on the site. In this factory, despite the high degree of mechanization, it was planned to employ some 500 workers.[85]

The main product of this production line was the basic universal panel of the system: 8 ft high and 3′4″ wide, wood framed and stressed skinned, with doors or windows where required, which was both structure and enclosure. This modular panel, which gave the scheme its flexibility, was an elegant but wasteful element, with a high degree of structural redundancy in the timber frame, and of a size which necessitated special 40″ sheets of plywood instead of the standard 48″.[86] The floor, roof, and ceiling panels were long, single-sided units, with triangular panelized gables and roof trusses. Later, to save costs in production, these were replaced by conventional 6″ × 2″ joists with glue-nail plywood on one side.[87] The structural parts, such as columns and sills, were moulded from solid lumber of excellent quality, bought rough and milled down, an expensive procedure. Filler strips, comprising narrow battenlike members, served to make up one, two, or three panel combinations, where they substituted for the missing panels of the universal four-member connection.[88] These were the component elements of the kit parts, out of which ideally any rectilinear design based on a 3′4″ module could be achieved.

Conceptually, this is where the design process ended: the theory of the packaged house envisaged a universal set of components out of which an almost unlimited number of alternative designs could be generated. It was hoped originally to market these components, this kit of parts, rather than complete houses. The realities of the situation, however, the exigencies of building regulations, FHA requirements for approval, the need to obtain mortgage financing, and the method of distribution—not directly to the owner—forced General Panel away from this dream of universality into the more conventional strategy of producing a limited range of standard houses. As we have seen, the initial production of General Panel was based on one house type only, the low-cost four-roomed model B-16. From the outset of the Packaged House project, despite the universality of the concept, the translation of the system into specific building types had been examined, sometimes in the greatest detail. We have already referred to several of these, commencing with the initial experimental house in Somerville, Massachusetts, based on the NHA's type TDU-1. In 1942, under Gropius, designs for standard house types A (a two-family house) and B (a continuous row house) were prepared; Wachsmann later developed the row house in much further detail. In 1943

9.15
General Panel, California, components being shipped by truck from factory, c. 1947

9.16
General Panel, California, sets of panels and other components, c. 1947

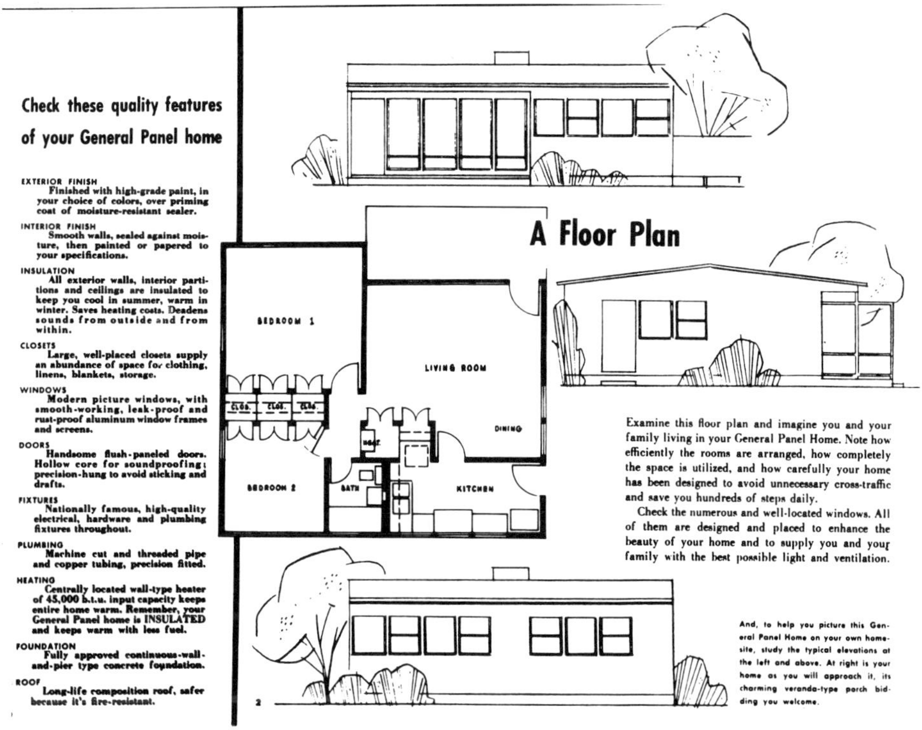

9.17
General Panel, page from advertising brochure, n.d.

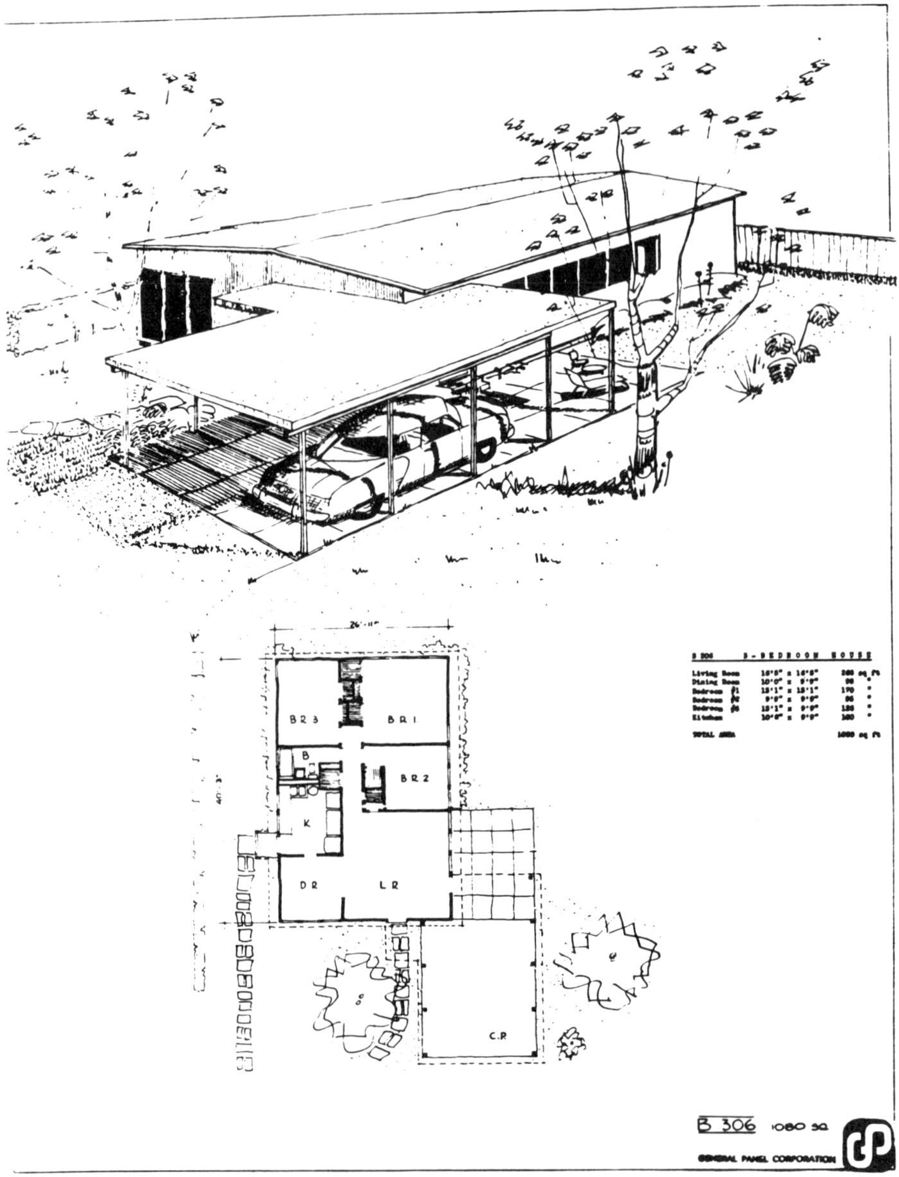

9.18
General Panel, California, standard house type B 306, n.d.

we have the proposals for the temporary dwelling unit and the field barracks. The following year Wachsmann designed and documented a two-bedroom cottage, while Gropius designed the flexible "growing and shrinking house" using General Panel modular elements. We have also noted various houses designed for the system by other architects, such as Neutra.[89]

By 1947, when the factory was ready to start production, a wide range of possible house types had been considered; some had even been illlustrated in the advertising brochures issued to the public.[90] Shapiro, an architect who did much designing and drafting for Wachsmann, presumably on a free-lance basis, had proposed preparing a folio of existing house types in June 1947;[91] whether this was done is not known, but there is in existence a set of loose-leaf pamphlets, illustrating 13 different General Panel models of one- to four-bedroom houses.[92]

While the house types were being prepared, and the factory equipped for mass production, a dealer network was being set up by the Celotex Corporation, in terms of its agreement with General Panel. In principle, an eclectic policy was adopted.[93] Franchise holders, who could be combinations of builders with other interested parties, such as real estate developers, would have exclusive rights within the franchise area. Prerequisites for a dealership were few but pragmatic: a capital of at least $50,000; a forklift truck with a gin pole rig; an undertaking to buy a new demonstration model each year; and finally, the drive of a promotion-minded salesman. They would receive training about the principles of the system and its advantages, and the construction manager would be trained in erection techniques by General Panel—despite the simplicity of the connector system, it was no "do-it-yourself" procedure.[94] In addition to these dealer-builder organizations, all generally small-scale operators, it was hoped to sell directly to the large speculative builders and real estate developers and of course to government agencies. To all outward appearances, it seemed that the dream of the factory-made house was, for Gropius and Wachsmann, at last about to be realized.

10

End of a Venture, End of a Dream?

General Panel: The Final Phase

During 1947 General Panel, embarking on the great adventure of the Burbank factory, sought to present to the outside world a picture that stressed its positive achievements: the excellence of the design, the high standard of the production plant. Two major articles in the national architectural press, the *Architectural Forum* and *Arts and Architecture*, reinforced this favorable image and painted a highly encouraging picture of the immediate future and long-term prospects of General Panel.[1] Attracted by the potential of General Panel, as it was perceived, and by the reputation of Gropius and Wachsmann, a press of highly qualified individuals, professionals who had formerly held office in the National Housing Administration, and architects with interesting backgrounds (one from the Bauhaus, another who had worked with Buckminster Fuller) offered their services to General Panel.[2]

But the optimistic picture was far from accurate. Although from time to time there had been articles probing the difficulties of the prefabrication industry,[3] yet in discussing the work of specific designers, especially architects as celebrated as Gropius or Wachsmann, they showed a reluctance to go below the surface of the problem, to enter too deeply into fundamental issues. In presenting a review of General Panel, they relied too heavily on material provided and too little on an independent critical analysis. Moreover they failed to relate the specifics of General Panel to the broader issues, and especially to the question of economic constraints within which the prefabrication industry operated. In their understandable attempts to do justice to the obvious merits of the Packaged House, they obscured the fact that all was far from well within the organization.[4]

In part the problems were personal, generated by the division of responsibility resulting from the establishment of the new company, and the ambiguity of Wachsmann's continued centrality to the project. He no longer had overriding control and differed strongly on several policy issues with the California directorate. As a case in point, the expedient of producing conventional floors and roofs, though undoubtedly making good sense economically, even technically, and intended only as an interim measure, caused Wachsmann great distress, as it detracted from that universality of the system which was, in a sense, its ideological justification.[5] Then, he was interested in exploiting new materials and methods (metal framing, plastic cores) which technologically had become much more exciting for him,[6] but although Southwell also advocated a shift to steel framing for the panels as a long-term move, the factory was irrevocably committed to wood. Wachsmann felt constrained and frustrated by what he saw as shortsightedness on the part of the administrators. They, in turn, though recognizing his undoubted genius,[7] were impatient of his eagerness to step beyond the realities of the day. Wachsmann was always willing to supersede his own inventions with something newer and better; the rigors of economic necessity committed the California factory to the status quo of the system as originally designed.

Indeed, it was the economic situation that now pressed most heavily on General Panel. Personal friction, philosophical disputes, could perhaps, with goodwill, be overcome. But the financial situation appeared threatening, almost desperate. The initial private investment and subsequent share issues, the mortgage from the War Assets Administration, the Reconstruction Finance Corporation's loan, the sum total of these funds proved completely inadequate for the development and equipment of the lavish Burbank factory. General Panel had been one of the few of the larger prefabricating firms to have raised capital through public stock subscriptions and loans. As Burnham Kelly pointed out, of the better-known prefabricators, "many owed either their original formation or much of their capitalization to large industrial empires."[8] General Panel had no such backing, and without these large reserves of capital, production was held up throughout the second half of 1947 by lack of working funds. This difficulty in securing the necessary financing, it was later explained, resulted from "the expiration of insured manufacturing financing provisions of the National Housing Act."[9]

The problem was twofold. The RFC loan was essentially short term and fell due, as Albert Wohlstetter explained, "before any substantial amount of manufacturing and constructing had been performed, and there was no means of repaying this debt from sales."[10] In addition the post-Wyatt interpretation that the marketing guarantee was valid for one year only failed to take into account the time necessary, in Wachsmann's view, "to buy carloads of material, produce, sell, ship and erect 10,000 houses."[11] This complaint may have been justified, in principle; Gropius, more sober and realistic in his judgments, also expressed disappointment at the failure of the government bureaucracy to understand the nature of the processes of industrialized building and how they differed from conventional methods in timing and financing. "The General Panel Corporation is somewhat stuck," he wrote in frustration to friends in Europe, "since the authorities are too backward and not geared for prefabrication. For that reason the financing of houses is very difficult to overcome."[12] Nevertheless, there were obviously other problems. It may have been impossible to produce 10,000 houses in the time available; but when Gropius was writing this letter, early in 1948, only fifteen houses had in fact been produced and sold, with a large number of incomplete sets of assorted panels also in stock.[13] Among reasons for production delays were technical problems, due to the fine tolerances and high accuracy demanded by the system, and to running-in difficulties generated by the "rush to production." Had time and circumstances permitted, it would obviously have been desirable to test the design and system in a pilot plant of limited production, before the final decisions on tooling the complete factory were taken.[14] Sales were also going slowly, due not to resistance to the product per se, but rather to its high cost. This went against all predictions of the great savings to be made (15 percent "across the board") by industrial production. But the hard fact remained that, until full production was achieved, materials costs were out of all proportion and the overheads (deriving from the expensive plant installation) prohibitive. Even Southwell's "worst case" assumption, that initial production, until full volume was achieved, would have to be subsidized by the corporation and sold at cost, proved overoptimistic.[15]

The financial crisis came to a head at the end of 1947. To avoid bankruptcy, there was a complete reorganization of the board of directors and officers of the corporation, in February 1948.[16] Carl Dahlberg resigned as president, and withdrew altogether from the affairs of the General Panel. Wendell had earlier resigned, and now Southwell also withdrew, effectively ending the direct Celotex connection. Albert Wohlstetter took over as president and general manager. He brought to his task a rich and varied background of industrial experience;[17] in addition he had for five years been associated with the prefabrication industry, not only through his connection with General Panel of New York since 1944 but, indirectly, through his position as program director of the NHA.

Konrad Wachsmann moved to California, on leave from the New York firm, and became vice-president. He was joined in this rank by Paul M. Fisher, who had served on the War Production Board (dealing with priorities in construction projects) and had then been chief of the Lumber and Plywood Section of the National Housing Agency, before joining General Panel in June 1947 as director of production and material control. With this reorganization of the administration of the corporation, a committee of creditors, who had anxiously met in January 1948, extended an opportunity for a recovery operation, by agreeing to a moratorium on debt repayment until 1 April 1948.[18] Wohlstetter responded by undertaking to try and renegotiate the RFC debt repayment, as well as to reschedule the payments to the principal suppliers, and by this means to shore up the shaky financial structure, while looking again at the problems of production and marketing.

In February 1948 production was diversified "to include contract milling, door production, manufacture of movie flats and a variety of other wood products."[19] There is, we suggest, an irony in this improvisation in the face of crisis, in a factory planned as a model of integration between product and production system. Production of the houses resumed, and the volume of production increased during 1948; but despite optimistic reports of "a large backlog of orders,"[20] it never reached significant proportions. Nor did the export market materialize, although such a development had always been actively considered.

We will recall that, in 1945, books of data had been prepared for Sweden and Canada and, in 1947, there had been high hopes of establishing assembly plants in Honolulu and Mexico City.[21] Later in 1947, prior to the establishment of the State of Israel, there had been approaches from Palestine, where the Near Orient Company, perhaps recalling the episode of the Palestine prefabs of the 1930s, explored the possibility of importing the General Panel components, on the basis of joint production in Palestine, in association with the Jewish Agency.[22] Overseas, after the war, this was a time of reconstruction, of resettling, of rehousing. From Germany, in 1948, a contact of Gropius suggested that the General Panel house could play a part in the reconstruction of the ravaged land of his birth.[23] But these approaches, tantalizing in the prospect they held of an ultimate breakthrough, all came to nothing.

Throughout 1948 and into 1949, the struggle for survival continued, and the financial problems multiplied. A scattering of houses were sold in California and Arizona. Associates of the company loyally supported it. Rudy Wolf, the chief draftsman, moved from New York to Los Angeles, and erected a General Panel house for his own use, which he occupies to this day; Nathan Mendelsohn, assistant treasurer, put up five in Riverside; and Josef van der Kar, an architect, who took over after Wachsmann left as technical director in January 1949 and remained in this capacity until December 1951, built a house using the General Panel system for a client in Pacoima.[24] After an excellent report on the quality of the General Panel house by the Corps of Engineers, the army erected a number of prefabs in Santa Monica, California, and shipped a special group with extra insulation and knockdown roof trusses to Alaska.[25] But the total number of all General Panel houses produced probably did not exceed 150 to 200, and financial limitations inhibited the acceptance of new business in an unbreakable vicious circle.[26]

10.1
Rudy Wolf, house for himself, Los Angeles, using General Panel components, c. 1948 (photographed in 1981)

Many of the problems that had initially troubled General Panel, problems with the unions, with local acceptance, were gradually solved. FHA structural approval was extended to cities across the United States, eventually to New York early in 1949.[27] But all these positive developments could not rescue General Panel, chronically underfinanced and unable to generate a market sufficiently large to justify the investment already made in its development and plant, still less the vast further investments required to put it on its feet.[28] Debts increased, tax liens were issued, suits were filed, eventually Wachsmann's precious machines themselves were mortgaged. For all practical purposes, by 1950, the end of the road was reached.

Konrad Wachsmann's own connections with General Panel Corporation were severed at this time. He left the California firm in 1948 or 1949, and his contract of employment by General Panel of New York, by which he had received a handsome weekly salary in addition to possible royalties on sales, expired on 11 September 1949, on the same day as his legal obligation to reveal to the corporation any improvements he might devise to the Packaged House system.[29] In a technical sense he was now free of his obligations; free, for instance, to answer a call by the Institute of Design in Chicago to head their Division of Advanced Building Research and start a new creative career. But in a real sense he was, in Gropius' bitter words, "brutally squeezed out,"[30] if not by his business associates, then by the harsh realities of the business world. Gropius, himself, had long ceased to play an active role in the corporation. He and Wachsmann retained their stockholdings, but by 1950 these had lost all real value; their loss was much more than a financial loss, it was the final dissolution of a long-cherished dream.

The General Panel Corporation went into liquidation probably at the end of 1951. The California corporation was suspended by the Department of the Secretary of State, California, on 1 July 1952, "for not complying with statutory requirements"; and the New York corporation was dissolved by proclamation on 15 December 1952, "for non-payment of franchise taxes." The Packaged House saga had come to an end: it lived on, however, in architectural myth and legend.

Failure of the Packaged House Venture

As a commercial venture the Packaged House of the General Panel Corporation was a resounding failure. A decade of dedicated work, an investment of $6 million, resulted in a trifling number of houses being produced. The causes of this failure were not, in any substantial degree, in the architectural conception, nor in the technological means, nor in the translation by industry of that conception into the reality of building components. Technically, according to both internal and objective external assessments, the Packaged House was a first-class product. Architecturally, it generated housing solutions which were, within reasonable limits, aesthetically acceptable to the prevailing taste and which were functionally more than adequate. The direct causes of failure lay elsewhere. Some of the causes were specific to General Panel; many were generic to the whole movement for the industrial production of houses and combined to doom many a promising venture to failure. It must be stressed that there is no single cause, no simple formulation, but in all cases complex interactions of many factors often cumulative in effect. This is especially true when we are dealing, not with the subsidized mass housing of Europe but with the production of that most emotionally charged, personal of possessions, the one-family house.

The inherent promise, in industrialized building, of significant cost reduction, in which Gropius and Wachsmann so firmly believed, was obviously only realizable in terms of mass production. It was on this premise that the Packaged House was conceived and the Burbank factory designed. But the problem was to reach this critical mass. In the late 1940s, as now, the United States was no socialist society, no welfare state, with a directed and subsidized program of mass housing, but a vigorous and highly competitive free-enterprise system. The prefabricator of houses in such a system could never begin to deal with the housing problems of the urban poor. Instead, he directed his attention to working- and middle-class housing in suburbia. He could try to provide for the less well-off, if not the really poor, through the provision of a minimum product, sited with little concession to space and amenity, at a very low cost; initially, at least, this was the preferred market sector of the mobile-homes manufacturer. It was a neglected sector in which neither the major developers nor the architects were active. Alternatively, the maker of prefabs could aim directly at the more ambitious, but hardly affluent, lower strata of the middle-class market, hoping to produce a factory-built product competitive in quality with the traditionally built houses of the tract developer, at a significantly lower price; this was the goal of General Panel, as it was of most other prefab firms. In this formidable task, where the high costs of research, development, and tooling could only be offset by large-scale production, the advocates of the factory-built house turned again and again to the paradigm of the automobile for encouragement and for justification. But this analogy was a false one. Car prices initially were high, to cover high tooling costs and disproportionate overheads, while production slowly increased. But as a generic product the car was unique, and its manufacturers had a complete monopoly; one either paid the high price or did not acquire a car. Eventually, of course, production rose to levels where prices could significantly be reduced, generating even larger demand. In more recent times one could see a parallel in the manufacture and marketing of computers. But industrialized housing did not produce a unique product, the competition of the traditionally built house was an ever-present factor, and the industry was denied that sheltered growth period it needed to reach the critical level of mass production. Wartime conditions blurred this essential truth for a short while, as did the massive support to the prefabrication industry given by the Wyatt program of the immediate postwar era. Yet when the General Panel house, after its long gestation, was eventually ready for production, even this temporary favorable condition no longer held.

In a sense the very high quality of the product contained within it the seeds of failure. What an irony is here! Held up by constant redesigning, in a search for an ever-better technical solution; by a perfectionist attitude to documentation; by the perceived need for the long-term testing of experimental models, one at least of which was laboriously handcrafted; by the determination to have a completely equipped model factory instead of a compromise ad hoc solution of improvised manufacture: delayed at each stage by this search for the ideal, it took much too long to move from initial concept to the final stage of actual production. This long drawn-out process proved costly in time and resources, and its consequences were disastrous.

The preliminary stages of design, and then the elaborate detailed follow-up in both the product itself and the production system, ate up vast amounts of capital (half a million dollars on research and development alone) and left both New York and California without financial resources when the critical production phase started. The long years of unprofitability deterred massive investment, and without adequate production capital the venture collapsed. Of course it could be argued that the scale of investment required was never really realizable for an independent company and that the whole venture was therefore in any event foredoomed to failure, "a gallant but quixotic enterprise," as Wohlstetter characterized it. This was true of normal times, but the housing crisis of the war and the immediate postwar period created abnormal conditions, where government subsidies and the unprecedented volume of demand were sufficient for those able to supply, to break through the constraints of limited venture capital. But because General Panel was not ready when the call came, because production only got underway, in a limited manner, in July 1947, the wartime boom was missed completely, and the postwar phase was caught too late, when the situation was rapidly reverting from crisis to normality.

The traditional housing industry, by 1947–48, was coping more than adequately with the housing demand, at least in the medium-price, middle-class market. It had learned to operate with considerable efficiency, incorporating many pre-made elements, utilizing much-improved methods of site mechanization, and organizing its flow of labor and materials with precision. In its on-site operations, at least in California, it was not penalized by inclement weather; its overhead costs were relatively low; and it could adapt to the vicissitudes of the market, in terms of numbers and in terms of architectural style, with more flexibility than could any production-belt system. The General Panel Corporation, on the other hand, was, from the point of view of production, much less responsive to market fluctuations and change, and there is doubt if it would have made economic sense even if it was able to exploit its full productive capacity, working three shifts a day, to make 30,000 houses a year.

At a more modest level of production General Panel Corporation could certainly not effectively compete with the houses of the building industry. Its houses were well designed and well made, using only the best materials, demanding precision in manufacture and erection, providing high levels of environmental performance, and with a structural redundancy inherent in a universal system: in other words, at low production levels, they were costly. But even at maximum levels of production, the factory-made house, where it was conceived, as was General Panel, as a closed system, was not likely to be cost competitive. Tract housing of a conventional nature could benefit from the economies of scale inherent in the mass production of nationally or regionally distributed building products and components; it could also benefit from the competitive nature of the building materials industry. Prefabricated houses predicated on open systems could also participate, if to a more limited degree, in these benefits. But a closed system such as General Panel could not freely incorporate elements from the competitive open market, and for all of its major components, its economies were limited to the scale of its own production. In other words, even with increased production, General Panel houses would probably not have been cheaper, to any notable degree, than conventional housing. In the climate of opinion that prevailed in the United States in 1948, a prefab house which was not demonstrably cheaper than its traditional competitor (as, for instance, the mobile home was to become) had little chance of capturing a significant segment of the middle-class housing market. This was not only an economic fact but a psychological one. A prefab such as the General Panel house might be of higher structural and performance standards than its conventional competitor, but it was perceived as inferior, because of the "prefab" tag, by the buying public. Similarly, although the General Panel house offered a real degree of flexibility in response to user needs unheard of in tract housing, it was perceived as being standardized and stereotyped. The term "prefab" had a pejorative connotation, which had to be discounted in the sale price. This was the bitter lesson which should have been learned by Gropius in his experience with the Hirsch copper houses in 1932.

But, in the final analysis, lack of public response was not the decisive factor. It is instructive in this respect to compare the failure of General Panel with the collapse of that other venture in high-technology house-manufacture, Lustron Homes. The Lustron steel-paneled house was reputed to be popular with the public and, despite increasing prices, sold well once production got properly underway.[31] Lustron Homes was an ambitious undertaking initiated and directed by Carl Stradlund of the Chicago Vitreous Enamel Co. Originally set up before the war, it finally acquired a war-surplus factory, the Curtis-Wright aircraft factory near Columbus, Ohio, in 1946, in the program initiated by Wyatt, who, until his exclusion from government housing matters, backed the company strenuously. They were financed by a prodigious RFC short-term loan, which by 1949 had been extended by the government to an astronomical $32.5 million. The scale of support is larger, and the projected production more extensive (a planned output of 100 houses a day), but the parallels with General Panel are obvious. Like General Panel, Lustron had a long gestation period and a late postwar start, and like General Panel, Lustron had a fully mechanized, highly complex, and very expensive production system. The reasons adduced for the failure of Lustron in 1950–51 are an echo of the problems bedeviling General Panel: high production costs leading to escalating prices; difficulties with organized labor; problems generated by the diversity of local building codes and a lack of sympathetic understanding in their application; the necessity of devising an adequate and properly financed distribution system; the unsuitability of bank mortgaging procedures when applied to an "instant" product rather than an extended building process; insuperable difficulties in raising sufficient capital, which short-term loans failed to alleviate; and, ultimately, the lack of a guaranteed and continuous market of adequate volume inherent in a free-enterprise housing system. All this occurred in the case of Lustron as with General Panel, notwithstanding the quality of design and technical excellence of the product. Because of the extent of the administration's involvement with Lustron, its ultimate collapse was spectacular; it contributed to the undermining of confidence in other prefabrication firms and eventually overshadowed in public attention the demise of General Panel.

In the eventual failure of both firms, as in their successes, personal factors were at play. Genius, creative imagination, willpower, and drive have a personal dimension, with both positive and negative connotations. Our account of the rise and fall of General Panel has amply demonstrated this truism, even if we do not go as far as Emerson in declaring: "There is properly no History—only Biography."[32] Yet the evidence indicates more than the thrust, creative or obstructive, of powerful and highly motivated individuals. It also points inevitably to the wider elements, the suprapersonal factors, the web of political, economic, and social issues, characteristic of culture, time, and place. And over and above the objective matters, there is the all-influencing if intangible spirit of the times, the *Zeitgeist* to which both Gropius and Wachsmann were dedicated.

Prefabrication: The Turning Point of Housing?

In the transition from traditional building methods to industrialized building, Konrad Wachsmann saw a critical point in the evolution of *homo fabiens*. "In discarding many of our old ideas about building," he contended in 1961, "we have reached a turning point. The decisions about what constitutes the formative energies have been made and the principles that will guide the developing forward movement are now apparent."[33] Despite the fact that this was written ten years or more after the General Panel debacle, and despite the fact that no other proposal of Wachsmann for industrialized building had reached a practical consummation,[34] his certainty about the ultimate goal, and his chosen path toward it, remained unshaken. He was absolutely convinced that a "turning point in building" had been reached, and that this was a point of no return. He believed that not only he personally but the building industry was irrevocably committed, in the long run if not in the immediate future, to the path of industrialization.

As Wachsmann penned these essentially hopeful thoughts, a more somber conclusion was being reached in Great Britain. At the war's end, a sweeping and courageous program had been initiated to tackle the short-term housing crisis through industrialization of the building process. Now, by the early 1960s, that program had come to an end, and in various forums, governmental and professional, it was being assessed. R. B. White, a British historian of prefabrication, gives the consensus view:

Out of all the verbal confusion and the years of experience, there is perhaps one outstanding fact that emerges: the illusory character of the assumptions that lower costs would inevitably and immediately *follow the mass production of houses (or other buildings) through widespread standardization and prefabrication. Circumstances have not so far combined with sufficient benevolence and constancy to warrant these assumptions. So far as this country is concerned, if experiments in alternative methods have proved anything positively, it is rather that "traditional" methods in building (particularly in house building) still commands a wide measure of support. . . .*[35]

These conclusions were a somewhat reluctant obituary notice for a program that by the standards of Hirsch Copper, Lustron, or General Panel, even by the standards of eminently successful American prefabricators like National Homes,[36] had considerable achievements to its credit.

In this program three products were particularly prominent:[37] Uni-Seco structures, utilizing timber framing and asbestos cladding; the Arcon house, designed by a consortium of architects and comprising an open system based largely on a wide range of existing materials and components; and the most sophisticated and completely industrialized system of all, the Aluminium Bungalow. This latter house was the result of the initiative of AIROH (Aircraft Industries and Research Organization on Housing), an organization sponsored jointly by the British Ministry of Aircraft Production and the aircraft industry to exploit the by then redundant capacity, in both manpower and industrial plant, of the wartime industries. The house was made in four segments, or modular units, each 7′6″ wide (to comply with road transport regulations), which were joined together on the site. At the heart of the house was a mass-produced service core comprising kitchen and bathroom. Despite minor technical defects which later developed—the eternal problems of corrosion and condensation—White was of the opinion that "the *Aluminium Bungalow* must be recorded as a great historical achievement in prefabrication."[38] This of course was a qualitative judgment of what was designed to be a house of limited life. Equally impressive was the quantitative performance: 29,000 Uni-Seco houses, 40,000 Arcon houses, and 55,000 Aluminium Bungalows, in the ten years of the program. To put this in perspective, we must recall that during the war, 104,826 family dwellings were built under the Lanham Act by all American prefabricators, and of these only 1,428 were regarded as permanent.[39]

10.2
Aluminium Bungalow, assembly line production, c. 1947

In Europe in the postwar period, the main thrust of construction was neither in the direction of these lightweight systems nor in low-density, single-unit, housing schemes. High-density housing developments were preferred, employing substantial construction systems, mainly in reinforced concrete. Throughout Europe, in the social-democrat welfare states of the west and the communist-dominated east, extensive state-sponsored housing programs made considerable use of precast concrete structural systems, prefabricated floors and story-height wall panels, and eventually complete boxlike housing modules. The technology was far more sophisticated, but the design and construction principles went back in essence to Ernst May's experiments in Stuttgart of the early twenties.

In the United States, in the 1950s and 60s, there was no such massive state intervention in the housing process, no assured and continuous market, and no large-scale development of comprehensive building systems. Nor did later governmental encouragement, as in the much-vaunted "Operation Breakthrough"[40] succeed in generating practical, economical, and viable industrialized building systems. There was, however, a lively theoretical interest, especially among architects and systems specialists, in the subject of systems building generally, and particularly in the development of prefabricated dwelling modules, the "new building block."[41] Wachsmann had little interest in these systems. He was looking, not for an economic solution to the housing problem, but to the elegant exploitation of advanced technology. He was drawn, philosophically, aesthetically (in the sense of a mathematician seeking the beauty of an elegant, minimalist, equation) to the materials of tomorrow, as revealed at the frontiers of metallurgical and chemical knowledge, rather than to the cumbersome and crude mass materials of yesterday; he was fascinated by the finesse of machine production, not by the quantitative bulk output of the concrete mixer. Gropius too did not embrace the "new building block," although he had not only pioneered the concept but had also, in his *Baukasten* project of 1923, given it its name. He saw in the repetition of large units, or of total dwellings, a perversion of technology, exploiting its mechanical potential through soulless multiplication of identical units, without the saving grace of variability and individual choice.

Gropius, more than half a century after his memorandum on prefabrication of 1910, returned to his original theme in an address at the Boston Architectural Center in February 1964.

Genuine variety without monotony could have been attained if we had taken greater interest and influence in the development and design of an ever more comprehensive production of standardized, component building parts *which could be assembled into a wide diversity of house types. Instead the idea of prefabrication was seized by manufacturing firms who came up with the stifling project of mass producing whole house types instead of component parts only. The resulting monotony further deepened the horror of a nostalgic, sentimental, unguided public of a prefabricated future.*[42]

Two points emerge from this notable address, which illuminate not only Gropius' philosophy but also his character. We are first reminded of the unswerving consistency of Gropius' thinking about prefabrication, for he here reiterates a theoretical position spelled out in 1910 and reinforced in a series of notable papers, as we earlier recounted, in the 1920s. And then once again we see the breadth of his vision, for his advocacy of prefabrication goes far beyond technical and economic considerations and embraces urban design, city form, and, eventually, what are for him the key issues: cultural integration and harmony in a world of diversity. Wachsmann recognized, and admired, this breadth of perspective in Gropius; the proper platform for the exercise of Gropius' extraordinary skills of synthesis, he later proclaimed, was as a minister of culture, or, within Unesco, as "cultural secretary general" where "he would have electrified the civilized world, the arts and probably even the sciences like a renaissance nobleman of his time. He would have done superb job."[43]

In the early 1960s, Gropius stood at a crossroads. Two vistas stretched before him, neither of which led to his desired goals. On the one hand, there was the road toward large-scale prefabrication, essentially practical only in terms of mass housing on the European model. In rejecting this approach, Gropius presents us with a paradox. Although he had been one of the principal advocates in Germany of high-density mass housing and had designed notable examples, he turned his back on this line of development when he migrated to the United States. In fact both he and Wachsmann at no time concerned themselves with the prefabrication of any form of large-scale housing blocks or dwelling clusters; instead, they remained faithfully wedded to the concept of the one-family house, made in component form in the factory. In this loyalty to the private house they were, despite their European background, more American than many native architects of their adopted land. Indeed, perhaps it was this very identification with the American way of life that they were trying, subconsciously, to prove.

The other road led to the production of the complete dwelling unit, generally utilizing lightweight systems. This form of prefabrication Gropius abjured on ideological grounds. In attacking it, his criticism was directed to two fronts: the developers like Levitt, mass-producing stereo-

10.3
Betonwerk Nedersachen, prefabricated heavy concrete construction, c. 1960

typed houses by on-site methods, and the manufactured house industry, embracing both the traditional prefabricators like National Homes, and the newly emergent mobile homes industry, both catering to the public taste by offering complete units differing only superficially, through what Gropius considered to be trivial stylistic frills.

It is interesting that neither Gropius nor Wachsmann ever seriously related to the mobile home industry, for it was to prove to be, according to informed judgment, "by far the most efficient building industry in the United States and probably in the world."[44] We can only assume their reasons for this lack of interest: Wachsmann because the mobile home, although fully industrialized, was technologically primitive and conceptually impure, thereby not challenging his creative abilities; Gropius, because the idea of a total product was abhorrent to his life's philosophy. And yet, in turning their back in indifference to the mobile home, they were ignoring the only really successful outcome of their search for the factory-made house. The reasons for the phenomenal success of this industry lay not only in the economy and utility of the product but in a much wider realm: the understanding that industrialized housing is not merely a technological system, but a total system. Arthur Bernhardt has spelled out this lesson in meticulously detailed analysis, in his important study of the industry. He summarizes his findings succinctly:

The mobile home industry has become the world's most efficient building industry because it has thoroughly and strategically manipulated virtually all the important functions that operate in or affect the larger building industry . . . . The fruit of the industry's labors is a comprehensive nationwide system, a total production–distribution–land development network effectively synchronized with its supporting and regulatory environments.[45]

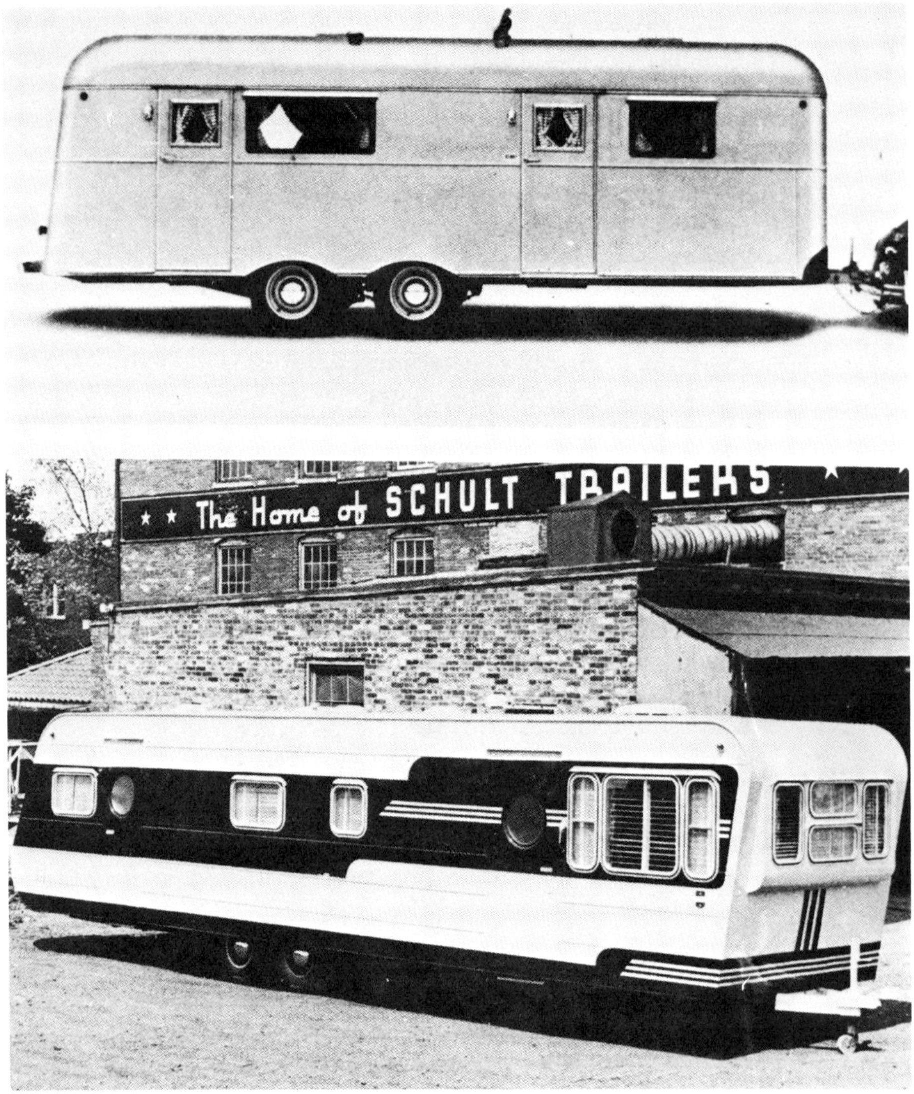

10.4
Mobile homes, 8-ft wide models.
Top: early 1940s; *bottom*: early
1950s.

Bernhardt uses two diagrams to illustrate this thesis. The one, entitled "Success Model," shows a complex system with all subsystems fully developed and in a balanced relationship; the other, the "Failure Model," shows neglect of several subsystems, and an overall imbalance in the total picture. We could adapt these diagrams as a pictorial representation of the Packaged House venture.

The success model, delineated in the balanced diagram of a total, complex, and integrated system, was the result of the foresight and pragmatic cooperation of many individuals and firms in the various branches of the mobile house industry. It was the outcome of a slow evolutionary process rather than the single masterstroke of a brilliant mind. A similar process did not take place in that branch of the prefabrication industry in which Gropius and Wachsmann, through General Panel, were involved. They could not be held to account for this. Their roles were essentially limited and specific. Wachsmann, naturally, devoted his energies to the technological base; and Gropius, from a very early stage, became detached from the broader policy issues, and his ability to influence policy formulation diminished, as he found himself on the sidelines of the affair. There is a deep irony in this, for if any man had, in principle, understood the necessity for an integration that went beyond technology, it was Walter Gropius. The idea of total synthesis, of a unity of art, industry, and society, was his life's philosophy, and the basis of his theoretical approach to every endeavor and, above all, to prefabrication. This he had made explicit in perhaps his most comprehensive theoretical statement, his "Systematische Vorarbeit für rationellen Wohnungsbau" of 1927, which we analyzed earlier, and in this amazing paper, with its emphasis on the varied and interlocking factors of industrialized building, he anticipated Bernhardt's major conclusions by more than 50 years. With the General Panel venture he failed, through his iso-

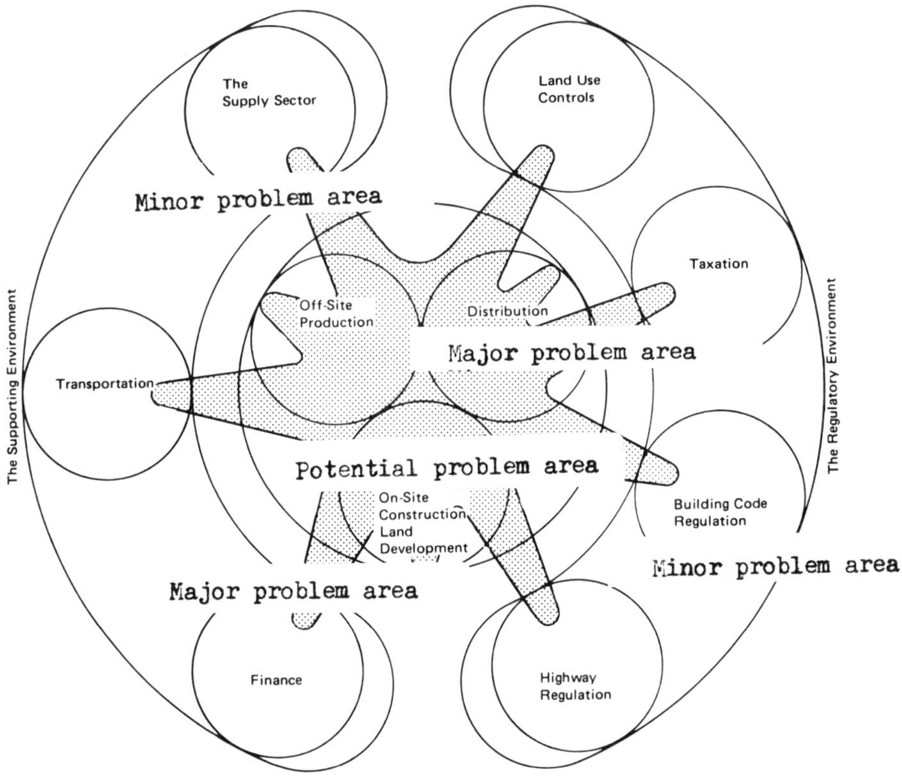

10.5
Bernhardt's scheme of industrial organization for homes production, adapted to show General Panel's areas of weakness

lation in Cambridge from the center of events and perhaps through a failure to understand the essence and intricacy of the American scene, to carry that theory into practice.

Both Walter Gropius and Konrad Wachsmann, in the years that followed the collapse of General Panel, went on to achieve great things in America: Gropius in the field of collaborative architectural practice; Wachsmann in the development of innovative structural technology, both of them as educationalists at the highest level, with an influence that continues to pervade creative thinking in architecture. Neither actively continued the struggle toward, nor saw the realization of their long-held dream of the house produced in its component form by industry, of high quality, low cost, and infinite adaptability to the changing needs of man. But they believed, to the end, that the dream was valid, awaiting the proper enabling conditions for its realization. These conditions embrace not only the proper tools for the comprehensive and rigorous analysis of the housing system, for as a complex system it is not given to the kind of intuitive analysis brought to bear even by such great men as Gropius and Wachsmann, but also a society more amenable to logical discourse, rational decision making, and creative human interaction than we at present appear to be. Yet, if the techniques and tactics of Gropius and Wachsmann were faulty, and their understanding of the intricacy of western industrial society limited, their basic insight into the ultimate necessity for industrialized housing seems valid. Gropius' optimism in this respect never diminished, despite the disappointments, despite the defeats, despite the many times he had "bloodied his nose"—to use his own, homely idiom[46]—in the conflict. "If your contribution has been vital, there will always be somebody to pick up where you left off, and that will be your claim to immortality."[47]

Appendix

Chronology

Walter Gropius	Konrad Wachsmann	General
1883 Born Berlin, Germany.		
		1887 Christoph and Unmack established in Niesky.
	1901 Born Frankfurt-Oder, Germany.	
1903–04 Technische Hochschule, Munich.		
1905–07 Technische Hochschule, Charlottenburg, Berlin.		1906 Hirsch-Kupfer established.
1907–10 Chief Assistant, Peter Behrens.		1907 Deutscher Werkbund.
1910 Memorandum to AEG on industrial production of houses.		1910 Le Corbusier in Behren's office.
1910–14 In private practice, Berlin.		

1914–18 Officer in German army.		1914–18 **First World War**.
		1914–21 Le Corbusier, mass-produced houses (Dom-Ino, Monol, Citrohan).
	1917–20 Apprenticed cabinetmaker and carpenter.	
1919 Founded Bauhaus, Weimar.		
	1920–22 Journeyman cabinetmaker and carpenter.	
1923 Baukasten-im-Grossen project.	1922–23 School of Applied Arts, Berlin.	1923 Le Corbusier, "Maisons en serie."
	1923–24 Academy of Arts, Dresden (under Tessenow).	1923 **Economic Chaos in Germany**.
	1924–26 Academy of Arts, Berlin (under Poelzig).	1924 Steel houses in Britain (Weir, Atholl, Telford).
		1924 Frigyes Förster, patent for metal-faced panel system.
1925–26 Bauhaus to Dessau.		1925 Taut, Wagner. Industrialization of the site, Berlin-Britz.
1926 Muche-Paulick steel house, Bauhaus.		1926 Steel houses in Germany (Wöhr, Böhler, etc.).
1926 "Wie bauen wir billigere, bessere, und schönere Wohnungen?"		1926 First prefab system in reinforced concrete for housing in Germany (Bron).
1926–28 Industrialization of the site, Törten-Dessau housing.	1926 Goes to work at Christoph and Unmack.	1926–29 May, prefabrication of the Siedlung, Frankfurt.

1927 Prefabricated houses at Weissenhof Siedlung, Stuttgart. ''Systematische Vorarbeit für rationellen Wohnungsbau.'' ''Der Architekt als Organisator der modernen Bauwirtschaft . . . .''	1927–29 Head of design studio, Christoph and Unmack, many projects.	1927 Buckminster Fuller, Dymaxion house.
1928 Resigns from Bauhaus, commences private practice in Berlin.		
1928 Business connections with Sommerfeld, visit to America.		
1929 Project houses for dry assembly (*Montagehäuser*).	1929–32 In private practice, Berlin, house for Einstein.	1929–32 **World Economic Depression**.
	1930 Publishes *Holzhausbau.*	1930 Hirsch-Kupfer acquires Förster-Krafft patent.
1930 Judge, competition, low-cost housing.	1930 Prizewinner, competition, low-cost housing.	1930 Bauwelt low-cost housing competition.
1931 Appointed architectural consultant, Hirsch Copper House Department, responsible for development.	1931 ''Kleine und grosse Bauten in neuer Holzbautechnik.''	1931 Building exhibition, Berlin. Prefabs exhibited (Hirsch copper house, Böhler steel house).
1931 Joins Growing House competition task force.		1931 Wagner organizes Growing House competition.
1932 Copper prefabs for Growing House exhibition, using Hirsch system.	1932 Wins Prix-de-Rome, goes to Italy, German Academy.	1932 Growing House exhibition, Berlin.

1932 Bauhaus transferred from Dessau to Berlin.		1932 Steel houses in the United States (Rowley, American Rolling Mills, General Houses, Inc., American Houses, Inc.).
1933 Bauhaus, Berlin, forced to close down.	1933 Compelled to leave Academy, now under Nazi control. Goes to Spain.	1933 **Hitler Comes to Power**.
1933 Prefab project, A. Rosa Works, Barcelona.	1933 Assistant to City Planner, Granada.	1933–38 Export of German prefabs to Palestine under transfer agreement (Deutsche Kupferhaus, formerly Hirsch; Böhler; Christoph and Unmack).
1934 Visit to Spalato. Meets Wachsmann. Visits Venice and Rome with Wachsmann. Goes to England.	1934 Visit to Spalato. Meets Gropius. Visits Venice; returns to Rome with Gropius.	
1934–37 In private practice, London, with Maxwell Fry.	1934–38 In private practice, Rome.	1934–39 Exodus of German architects, to Britain, United States, Palestine.
1937 Moves to USA. Professor at Harvard.		1937 The Wagner (U.S. Housing) Act.
1937–41 In practice with Marcel Breuer.		
1938 Chairman, Department of Architecture, Harvard.	1938 Moves to France. First proposals for Mobilar system.	
	1939–40 Interned as enemy alien in France. First proposals for panelized system of construction for prefabs.	1939–45 **Second World War**.

1940 Uses his influence to get Wachsmann released from internment.	1940 Released from internment, short period of service in French army.	1940 The Lanham Act.
1941 "How to Bring Forth an Ideal Solution to the Defense Housing Problem," with Martin Wagner.	1941 Visa obtained, sails for the United States.	
1941 Invites Wachsmann to stay at Gropius home in Lincoln.	1941 Travels to Lincoln. Works on projects with Gropius.	
1941–42 Design of Packaged House system with Wachsmann.	1941–42 Design of Packaged House system with Gropius.	1941 **USA Enters War**.
	1942 Moves to New York.	
	1942 General Panel Corporation, New York, established.	1942 All federal housing functions consolidated in NHA.
1943 Organizes demonstration of General Panel houses.		
	1944 Office partition system for General Panel.	
1945–69 Partner in The Architects Collaborative.	1945 Mobilar hangar system.	1945 **End of War**.
	1946 General Panel Corporation, California, established.	1946 Veterans Emergency Housing Program.
	1949 Contract with General Panel Corporation expires. Appointed Professor, Institute of Design, Illinois Institute of Technology, Chicago.	

	1950 General Panel ceases to function.	1950 **Korean War**.
1952 Professor Emeritus, Harvard.		
	1959 "Wendepunkt in Bauen."	
	1965 Professor, School of Architecture, University of Southern California, graduate program in industrialization.	
1969 Died Boston.		
	1980 Died Los Angeles.	

Notes

Introduction

1. From "The Hollow Men," in T. S. Eliot, *Selected Poems*, Penguin Books, Middlesex, 1948.

Chapter One

1. Richard Bender, *A Crack in the Rear-View Mirror*, 1973, p. 6.

2. According to W. Hebebrand, *Zur neuen Stadt*, Berlin, 1969. (cited in Karin Wilhelm, "From the Fantastic to Fantasy," *Architectural Association Quarterly*, 11:1, 1979, p. 15, n. 22), Giedion had thus described Oud's housing at the Kiefhoek estate.

3. From the catalog of C. D. Young & Co., *Illustrations of Iron Structures*, c. 1856.

4. John Burchard, in Bemis, *The Evolving House*, 1936, p. 335.

5. Buckminster Fuller, "Dymaxion Houses," *Architectural Record*, 75, January 1934, p. 10.

6. Gilbert Herbert, "The Architectural Design Process," *British Journal of Aesthetics*, 6:2, April 1966. Attention is drawn to Louis Kahn's poignant: "I dream of spaces full of wonder . . . when I place the first line on paper to capture the dream, the dream becomes less" (p. 162). The concept of concretization of the design process is Esherick's: "The real problem lies in working our way back—by design—to the real world" (p. 162).

7. The emergence of the idea of systems has been summarized in Gilbert Herbert, *Holism, the Ecosystem and Architecture: Towards a Philosophy of Architectural Design*, 1975.

8. This theory of architecture, in Gropius' thinking, was but an integral part of a much wider, holistic, world view. For an exposition of this philosophy, see Gilbert Herbert, *The Synthetic Vision of Walter Gropius*, 1959.

9. Marinetti's "Initial Manifesto of Futurism," 1909, celebrates the beauty of speed, the power of the machine, the dynamism of factories—but as symbols of the destruction of ancient values rather than a creative principle.

10. For an account of this development, in relation to Britain and its dependencies, see Gilbert Herbert, *Pioneers of Prefabrication*, 1978; in the United

States, see Charles E. Peterson, "Early American Prefabrication," *Gazette des Beaux-Arts*, 6:33, 1948; in general, Heinrich Wurm, "Vorgefertige Bauwerke des 19. Jahrhunderts," *Technikgeschichte Bd.*, 33:3, 1966.

11. Both Manning and Hemming of Bristol, two of the most important makers of buildings for export in the nineteenth century, have testified that their entry into the business of prefabrication was motivated by the wish to provide shelter for their sons and their friends, who had emigrated to Australia. See Herbert, *Pioneers*, 1978, pp. 4, 62.

12. According to Ronald Lewcock, *Early-Nineteenth-Century Architecture in South Africa*, 1963, p. 191.

13. Gilbert Herbert and Paul Stark, *Manning Houses: Their Provenance and Extent in South Australia*, 1979.

14. Charles E. Peterson, "Prefabs for the Prairies," *Journal of the Society of Architectural Historians*, November 1952; "Prefabs in the California Gold Rush, 1849," *Journal of the Society of Architectural Historians*, December 1965.

15. The prime minister was the South African General J. C. Smuts, whose home, Irene, was a converted corrugated iron British army prefab of the turn of the century (Herbert, *Pioneers*, 1978, pp. 145–147). Other eminent men who lived in prefabs ranged from King Eyambo, who lived on the Calabar River, in a palace made by Laycock of Liverpool (1843) to Lieut. Governor La Trobe, of Port Phillip, whose Manning Cottage (1839) still stands in Melbourne today.

16. Examples of such handsome cast-iron villas still in use today include Corio Villa, in Geelong (c. 1856), and Tintern, in Melbourne (c. 1855).

17. According to an advertisement in Burton, *Cape Colony for the Settler*, 1903.

18. The first published account of this significant experiment is in a letter from Benson et al., "Early Examples of Prefabrication," *Royal Institute of British Architects Journal*, June 1963, p. 251. A further, illustrated, description of Brodie's work is Moore, "An early system of large-panel building," *Royal Institute of British Architects Journal*, September 1969, pp. 383—386.

19. For accounts of the work of Grosvenor Atterbury, see John Burchard, in Bemis, *The Evolving House*, 1936, pp. 349–353; Bruce and Sandbank, *A History of Prefabrication*, 1945, pp. 30–33; and Burnham Kelly, *The Prefabrication of Houses*, 1951, pp. 12–13.

20. Ida Darlington, "Thompson Fecit," *Architectural Review*, September 1958, pp. 187–188.

21. Described in Fairbairn, *On the Application of Cast and Wrought Iron to Building Purposes*, pp. 165–176.

22. Ida Darlington, "Thompson Fecit," 1958, pp. 187–188.

23. Kelly, *The Prefabrication of Houses*, 1951, p. 15.

24. Kelly, *The Prefabrication of Houses*, 1951, pp. 15–16.

25. R. B. White, *Prefabrication: A History of its Development*, 1965.

26. White, *Prefabrication,* 1965, p. 80.

27. Kelly, *The Prefabrication of Houses,* 1951, p. 60.

28. White, *Prefabrication,* 1965, p. 79.

29. The analysis of apartment plans was carried out on a systematic basis by Alexander Klein for the German National Housing and Building Research Society and was published in 1928 (see Catherine Bauer Wurster, "The Social Front of Modern Architecture in the 1930s," *Journal of the Society of Architectural Historians,* 24:1, March 1965, p. 50). The minimal dwelling was examined by Bruno Taut, "Grenzen der Wohnungsverkleinerung," *Deutscher Bauzeitung,* 29 April 1931, pp. 210–212, and by Klein, "Die Kleinwohnung als Wirtschaftliches, Wohntechnisches und Raumgestalterisches Problem," *ibid.*, pp. 213–216. The concept of *Existenz-minimum,* essentially German, was published by CIAM in 1930 as "Der Wohnung für das Existenz-minimum" (according to Wurster, "The Social Front," p. 49). For early studies of flexibility in the dwelling plan, see Walter Bockman, "Die veranderbare Wohnung," *Bauwelt,* 12, 19 March 1931, pp. 394–395, for a proposal of 69 sq m and 90 sq m apartment shells, with fixed bathrooms and kitchens, and 12 variations of subdivision for each shell; and Otto Schmidt, "Verstellbare Wande für variable Wohnung," *Bauwelt,* 36, 3 September 1931, pp. 1156–1157, for details of adjustable partitions for flexible dwellings, with special reference to Bockman's apartment shells.

30. See, for instance, Ernst May, in *Das neue Frankfurt,* n.2–3, v.4, p. 36 (cited in Nicholas Bullock, "Housing in Frankfurt 1925 to 1931," *Architectural Review,* 163:976, 1978, p. 337): "The ideal and most natural form of housing is the detached house. . . . A dwelling in a multi-story tenement block can never create for the family, least of all for the children, that healthy way of living that is provided by the single house."

31. As an example, we may cite the low-cost houses exhibited at the Bauwelt-Musterschau of 1931, and the Häuser zu Festen Preise exhibition (as published in *Bauwelt,* 9, 3 March 1932, pp. 221–254).

Chapter Two

1. During the nineteenth century architects such as J. B. Snook, Thomas & Son, F. A. Petersen, King & Kellum, and J. F. Duckworth frequently used the Badger pre-made elements in their buildings (W. Knight Sturges, *The Origins of Cast Iron Architecture in America,* 1970); the ironfounder C. D. Young used the services of architects Bell and Miller, and Bellhouse's famous customs house in Peru was designed by the well-known architect E. Salomons (see Herbert, *Pioneers of Prefabrication,* 1978, pp. 159, 161, 296n, and 69).

2. Stanford Anderson, *Peter Behrens and the New Architecture of Germany, 1900–1917,* 1968, p. 205.

3. Anderson, *Peter Behrens,* 1968, p. 211.

4. This trenchant phrase remains in my mind, but the author escapes me!

5. But see the polemical battle between Van de Velde and Muthesius at the 1914 Werkbund Congress, over the issue of standardization. Anderson, *Peter Behrens*, 1968, pp. 376–381.

6. This discussion follows Anderson, *Peter Behrens*, 1968, ch.5.

7. Peter Behrens, *Elektrotechnischer Zeitschrift*, 31, cited in Anderson, *Peter Behrens*, 1968, pp. 202–203.

8. Harry Kessler, *Walter Rathenau*, Berlin, H. Klemm, 1928, p. 107, cited in Anderson, *Peter Behrens*, 1968, p. 196.

9. Walter Gropius, born Berlin, 18 May 1883, died United States, 1969.

10. Gropius mentions Walter Rathenau in his article "Der Gedanke der Rationalisierung in der Bauwirtschaft," 1929.

11. Nikolaus Pevsner, *Pioneers of Modern Design: From William Morris to Walter Gropius*, 1960.

12. Except for some marginal comments in "Gropius at Twenty-six," *Architectural Review*, July 1961, p. 49.

13. Gropius, "Programm zur Gründung einer allgemeine Hausbaugesellscahft auf künsterlische einheitlicher Grundlage, m.b.H.," typescript, dated March 1910. According to Hans Wingler, *Bauhaus*, 1969, p. 20, it was presented to Rathenau in April 1910.

14. Siegfried Giedion, *Walter Gropius: Work and Teamwork*, 1954, p. 74, and Pevsner, *Pioneers*, 1960, p. 38, both give the date of the memorandum as 1910, Pevsner basing this date on a letter from Gropius himself.

15. There seems to have been no history of conflict between Behrens and Gropius, despite Gropius' resignation over disagreements with Behrens on the Cuno house, in 1910 (for reports on this incident see Alan Windsor, *Peter Behrens*, 1981; and Anderson, *Peter Behrens*, 1968, p. 401). Behrens served on the Board of Governors of the Circle of Friends of the Bauhaus (see Wingler, *Bauhaus*, 1969, p. 78), and Gropius wrote with respect of Behrens (see Gropius, *Apollo in the Democracy: The Cultural Obligation of the Architect*, 1968, p. 1965).

16. Gropius, in a letter to Wulf Herzogenrath, 30 October 1968 (cited in Reginald Isaacs, *Walter Gropius: Der Mensch und sein Werk*, 1983, p. 93), wrote: "Selbstverständlich, da ich zwei Jahre im Büro von Behrens gewesen bin und viele Gespräche mit ihm geführt habe, ist es möglich, dass ich ihn um Rat fragte." The memorandum does not appear on any authoritative list of Behrens' work (such as in Buddensieg and Rogge, *Industriekultur: Peter Behrens und die AEG*, 1978). The fact that Behrens' name is not mentioned in the original document is not conclusive, for neither is Gropius'. When the present author, in 1955, suggested in a paper that the work was done "in collaboration with Behrens," reading perhaps too much into the statement by Giedion (*Walter Gropius*, 1954, p. 74, my emphasis) that the memorandum was prepared "while Gropius was working *with* Peter Behrens, a misinterpretation reinforced by my personal assessment that this was most likely, Gropius did not reject

this interpretation, and was even prepared to write the foreword to the book in which it eventually appeared (Herbert, *Synthetic Vision*, 1959).

17. Buddensieg, *Industriekultur*, 1978, pp. D112–D113.

18. Illustrated in *Architectural Review*, July 1961, p. 50.

19. Anderson, *Peter Behrens*, 1968, p. 346.

20. All quotations from "Gropius at Twenty-six," *Architectural Review*, July 1961, pp. 49–51, a complete English summary of the memorandum.

21. For a note on Edison's experiments in prefabrication, see Burchard, in Bemis, *The Evolving House*, 1936, pp. 410–412.

22. Macfarlane's contribution to prefabrication is discussed in Herbert, *Pioneers of Prefabrication*, 1978, pp. 173–186, passim.

23. As Franciscono, *Walter Gropius and the Creation of the Bauhaus in Weimar*, 1971, explains *Handwerk* means not handicraft but production in small factories.

24. Arnold Whittick (in *Encyclopaedia of Modern Architecture*, London, Thames and Hudson, 1963, p. 170), writes of Le Corbusier: "He was occupied from 1911 to 1912 with questions of mass-production and standardization in the circles of the Deutscher Werkbund."

25. In the English version, this chapter is called "Mass-Production Houses." I quote from the translation of Frederick Etchells, *Towards a New Architecture*, London, Architectural Press, 1927 (1946 edition), pp. 209–247.

26. Gropius left Behrens' office in mid-June 1910 and entered into private practice in Berlin with his friend Adolf Meyer, also a former assistant of Behrens.

27. Gropius founded the Staatliche Bauhaus in Weimar in April 1919. He moved with the Bauhaus to Dessau in 1925 and remained the director until his resignation in February 1928.

28. The slogan: "Art and technology: a new unity," had first been put forward by Gropius in a lecture at the time of the Bauhaus exhibition in the summer of 1923. The next year, he included it in the draft of his "Breviary for Bauhaus Members" (Wingler, *Bauhaus*, 1969, pp. 69, 76).

29. It was not only opposed by the artist Feininger but also by the more technically oriented Muche. (Wingler, *Bauhaus*, 1969, pp. 97, 113–114).

30. Wingler, *Bauhaus*, 1969, p. 111.

31. In a circular to Bauhaus Masters on 3 February 1922, Gropius wrote: "It is possible that the work in the Bauhaus workshops will lead more and more to the production of single prototypes." (Wingler, *Bauhaus*, 1969, p. 52).

32. Wingler, *Bauhaus*, 1969, pp. 109–110.

33. Wingler, *Bauhaus*, 1969, pp. 109–110.

34. Breuer, in 1925, designed unit furniture on the basis of a square modular grid of 33 cm. At no stage, however, was

this module related to the architectural spaces containing the furniture.

35. According to Pevsner, *Pioneers*, 1960, p. 35.

36. The quotation, with my emphasis added, is from Wingler, *Bauhaus*, 1969, p. 385. The products referred to were mainly in the form of furniture and equipment, but the external walls were of "Jurko" plates.

37. The drawings for the system had the following legend: "Lösung der gegensätzlichen Forderungen nach grösstmöglicher Typisierung (Wirtschaftlichkeit) und grösstmöglichter Varibilität der Wohngebäude." Gropius, *Apollo*, 1968, p. 83.

38. See Wolfgang Pehnt, "Gropius the Romantic," *Art Bulletin*, September 1971, pp. 379–392, for a discussion of this phase of Gropius' work.

39. The project was first reported by Gropius in "Der grosse Baukasten," *Das neue Frankfurt*, 1, 2, 1927–28, pp. 25–30. The definitive account is Gropius, "Versuchsiedlung in Dessau," *Reichsforschungsgesellschaft für Wirtschaftlichkeit in Wohnungsbau*, 7, 1929, pp. 1–136 (special number). Wingler gives a summary account in English in *Bauhaus*, 1969, pp. 126–127, 414–416, based in part on Gropius, *Bauhausbauten Dessau*, 1930.

40. This society was established in June 1927 in order to carry out research into sociological, economic, and technical aspects of housing. In addition to Gropius, architects such as Bartning and May were associated with it. For an account see Wiedenhoft, *Workers' Housing in Berlin in the 1920s*, 1971, p. 111.

41. Wingler, *Bauhaus*, 1969, p. 414.

42. Gropius himself differentiated between the manufacture of building elements erected on site by mechanical equipment, such as precast concrete panels, and pure factory-made systems ("Serienmässige, reinfabrikatorische Bau von Montagehäusern"). Gropius, "Der Architekt als Organisator der modernen Bauwirtschaft und seine Forderungen an die Industrie," 1927.

43. See discussion in D. C. Anderson, *Architecture as a Means for Social Change in Germany, 1918–33*, 1972, p. 56.

44. Cited in Wilhelm, "From the Fantastic to Fantasy," 1979, p. 7.

45. Discussed in Anderson, *Architecture as a Means for Social Change*, 1972, p. 92.

46. For an account of this scheme see Lion, "Die ersten Wohnungsbauten aus Betonplatten in Deutschland," *Deutsche Bauzeitung*, 1926, pp. 112–115.

47. See, for instance, Louis Perret's proposal, fully detailed and illustrated, for the design and erection of a prefabricated precast concrete multistory block of flats, including pipe ducts, alternate allocation of floor spaces, stages of construction, and an explication of the advantages of prefabrication as a building system, which was published in Louis Perret, "L'Immeuble normalisé a montage rapide," *Arts et*

metiers, 94, July 1928, pp. 241–246.

48. For an account of "Praunheim" see *Das neue Frankfurt*, 4–5, April–May 1930. See also Anderson, *Architecture*, 1972, p. 92; Wilhelm, "From the Fantastic to Fantasy," 1979, pp. 7–8; Kramer, "Das neue Frankfurt," *Architectural Association Quarterly*, 11:1, 1979, pp. 44–49; Bullock, "Housing in Frankfurt," 1978, pp. 335–342.

49. According to Anderson, *Architecture*, 1972, n. 172, the housing shortage in Germany in 1920 was 1 million dwellings. The situation steadily deteriorated until 1924, when the major effort at rehousing commenced. By 1930 (n.180) the cooperative housing movement alone had built 650,000 dwellings, housing 2,360,000 people.

50. This account is based on information in Stratmann, "Housing Policies in the Weimar Republic," *Architectural Association Quarterly*, 11:1, 1979, pp. 16–23; and Bullock, "Housing in Frankfurt," 1978, pp. 335–342.

51. In addition to Muche's own account in the journal *Bauhaus* (which has not been available to me), the steel house has been described and illustrated in "Deutsche und Österreichische Stahlhäuser," *Moderne Bauformen*, July 1927, pp. 5–11 passim; and in Spiegel, *Das Stahlhausbau*, 1928, pp. 92–94 (v.1).

52. For an account of Braune and Roth see Spiegel, *Das Stahlhausbau*, pp. 84–90. The first house by this firm was completed in August 1926.

53. According to "Deutsche . . . Stahlhäuser," 1927, p. 7.

54. This house was exhaustively discussed by Gropius, "Siedlung in Stuttgart, Weissenhof," *Reichforschungsgesellschaft für Wirtschaftlichkeit in Wohnungsbau*, 7, 1929, pp. 104–106, 133, 136, 144, 147, 148. In more accessible form it appears in Rasch, *Wie Bauen?* 1927, pp. 142–143; and the Deutscher Werkbund publication, *Bau und Wohnung*, 1927, pp. 59–67.

55. Gropius, "Wie bauen wir billigere, bessere, schönere Wohnungen?" manuscript, 1926/27.

56. Gropius, "Systematische Vorarbeit für rationellen Wohnungsbau," *Bauhaus*, 2, 1927; trans. in Wingler, *Bauhaus*, 1969, pp. 126–127.

57. Gropius, "Der Architekt als Organisator der modernen Bauwirtschaft und seine Forderungen an die Industrie", manuscript, 1927.

58. "Will Handwerk und Architektenschaft beiseite stehen, wenn sich heute bedeutende Gruppen, wie die Stahlverbände in Wirtschaftlichem zusammengehen entschlossen haben, das Land mit tausenden von Stahlhäusern zu beliefern, nach völlig veränderten Herstellungsmethoden?!" Gropius, "Der Architekt als Organisator," 1927, p. 6.

59. The following account of Gropius' philosophy is based in general on Herbert, *The Synthetic Vision of Walter Gropius*, 1957, especially chapter 2: "The Organic Nature of Design."

60. Gropius, *The New Architecture and the Bauhaus*, 1935, p. 27.

61. Gropius, *The Scope of Total Architecture*, 1955, p. 72.

Chapter Three

1. Hans Spiegel, *Das Stahlhausbau*, v. 1, 1928, pp. 122–124.

2. Department of Commerce, *Construction of Steel Houses Increasing in Western Germany*. Washington, Special Circular No. 703, Iron and Steel Division, 23 May 1928.

3. Kelly, *The Prefabrication of Houses*, 1951, p. 17.

4. Junkers, the aircraft manufacturer, developed the steel lamella roof truss, and experimented with stressed steel structures (ibid.; Spiegel, *Das Stahlhausbau*, 1928, pp. 12–13). The Stinnes industrial empire comprised not only coal mines and shipping but major interests in the iron and steel business. The enterprises of Stinnes, Thyssen, Phoenix AG, and Otto Wolf were linked together in the Vereinigte Stahlwerke trust, established in 1926 (see Thompson, *Europe since Napoleon*, 1966, p. 656).

5. According to the classification by Spiegel, *Das Stahlhausbau*, 1928, pp. 26–27, these terms were in common usage.

6. Department of Commerce, *Construction of Steel Houses*, 1928.

7. Both Neutra, *Wie Baut Amerika?* 1927, and Mendelsohn, *Russland, Europa, Amerika*, 1929, refer to the steel structures of America. Spiegel, *Das Stahlhausbau*, 1928, pp. 29ff., devotes a chapter to "Stahlhausbauten in Amerika" and pays special attention to the Tappan, Broderick, and McKay systems.

8. In the earliest reference noted so far to the British experience, Blecken, "Stahlhäuser," *Deutsche Bauzeitung*, 20, October 1926, p. 149, refers approvingly, and at length, to the official support given in Britain to steel house production. The Weir house is emphasized in Spiegel, *Das Stahlhausbau*, 1928, pp. 17ff., in the chapter "Stahlhausbauten in England und Frankreich." Weir is also discussed in Schmid, "Wohnhäuser nach dem Stahlwandsystem," *Zeitschrift des Österr. Ingenieur- und Architekten-vereines*, 19–20, 1927, p. 174; in "Deutsche und Österreichische Stahlhäuser," 1927, p. 5 (which refers to a previous article as "unserem Aufsatz über die englischen 'Weir-Häuser' "); and especially in Lion, "Stahl-Wohnhäuser," 17–18, 1927, pp. 156–158, which is devoted exclusively to a study of the Weir house.

9. On the derivation of the Wöhr house see Spiegel, *Das Stahlhausbau*, 1928, p. 83; "Ein Stahltafelbau nach dem Vorbild des englischen Weirhaus"; and Lion, "Stahl-Wohnhäuser", 1927, p. 158; "Die Firma Wöhr, Unterkochen, baut änliche wie die Hersteller der Weir-Häuser. . . ." Although the Wöhr house is the first postwar model we know of, Dex Harrison, *A Survey of Prefabrication*, 1945, p. 7, reports: "The first steel house we have recorded in Germany was the Scherrer house, brought out in 1915 by a firm which had previously produced prefabricated timber houses, and it was a load-bearing panel type. The extent of use of this system is not known, and sub-

sequent German steel systems developed out of the experience gained in Britain in the early 1920s, and especially from the Weir type."

10. For a discussion on the role of shipbuilders in early prefabrication see Herbert, *Pioneers*, 1978, pp. 41–42, 49–50.

11. Department of Commerce, *Construction of Steel Houses*, 1928.

12. Blecken, "Stahlhäuser," 1926, pp. 149–151.

13. Heinrich Blecken, "Neuzeitlicher Stahlhausbau," *Deutsche Bauzeitung*, 3, March 1929, pp. 31–35.

14. The name is variously spelled Schmid or Schmidt, sometimes in the same article!

15. Schmid, "Wohnhäuser nach dem Stahlwandsystem," 1927.

16. Lion, "Stahl-Wohnhäuser," 1927.

17. Spiegel, *Das Stahlhausbau*, 1928, pp. 97–102.

18. Department of Commerce, *Construction of Steel Houses*, 1928.

19. For a descriptive account of this method, see *Bauwelt*, 46, 17 November 1932, p. 1160.

20. Illustrated in *Bauwelt*, 9, 26 February 1931, p. 28.

21. A note to this effect appears in *Bauwelt*, 19, 17 May 1931, p. 664.

22. Illustrated in *Bauwelt*, 9, 3 March 1932, items 7, 30, 40, 62; and ibid., 21, 26 May 1932, p. 528. Architects using the Böhler system include Johannes Niemeyer, Hugo Virchow, Kuhnert and Pfeiffer and Fritz Glanz.

23. We learn this from a letter from Gropius to Böhler Co., 12 August 1931.

24. "The fine pioneer work of the Germans with metals and concrete," writes Burnham Kelly, of German prefabrication in the 1930s, " . . . was curtailed by the depression in 1932, and completely halted when the Nazis came to power in 1933. The use of steel for house-building was prohibited, and all civil building was limited by the channeling of resources to the construction of fortifications and military and party buildings." (Kelly, *The Prefabrication of Houses*, 1951, pp. 17–18). This date is open to question, and we shall defer judgment until we deal with other examples of Böhler's.

25. Burchard, in Bemis, *The Evolving House*, 1936, p. 367.

26. Adolf Sommerfeld, "Was will der 'Deutsche Holzbau-Verein'?" *Der Holzbau* (supplement to *Deutsche Bauzeitung*), 54:1, 1920, p. 1; "Marine-Normalflugzeug-Halle," *Der Holzbau*, 54:1, 1920, p. 2.

27. These buildings are discussed in Wolfgang Pehnt, "Gropius the Romantic," *Art Bulletin*, September 1971, p. 383.

28. Pehnt, "Gropius the Romantic," 1971, p. 384.

29. "Neuzeitliche Blockhaus-Bauweise," *Der Holzbau*, 54:6, 1920, pp. 21–23; Marine-Baurat Hahn, "Neuzeitliche Blockhausbauten," *Der Holzbau*, 54:21, 1920, pp. 81–84.

30. "Die Holzbau-Siedlungen des 'Wohnungs-Verbandes Gross-Berlin,' " *Der Holzbau,* 54:13, 1920, pp. 49–52.

31. Leo Kuhberg, "Die Holzbauweise in der technischen Zukunft Deutschlands," *Deutsche Bauzeitung,* 24, December 1926, pp. 121–123.

32. "Technische Einzelheiten auf der Berliner Sommerschau," 1932, pp. 590–591.

33. For Karl Schmidt, see Pevsner, *Pioneers,* 1960, pp. 34–35, p. 223 n.87; also Gillian Naylor, *The Arts and Crafts Movement,* 1980, pp. 184ff.

34. Pevsner, *Pioneers,* 1960, p. 34.

35. *De-We Holzhäuser,* 1929.

36. Riemerschmid, "Holzhäuser," *Bauwelt,* 40, 1932, pl. 5–8.

37. "60 billige zeitgemässe Eigenhäuser," *Bauwelt,* 9, February 1931, pl. 13.

38. Wachsmann, *Holzhausbau,* 1930, pp. 62–67.

39. Both Mrs. Wachsmann and I have been unable to elicit any information from Christoph and Unmack, who claim that all their records were destroyed during the war. This history of the firm is based on the following sources: Wachsmann's manuscript for his still to be published autobiography; Heinrich Wurm, "Die Industrialisierung des Holzhausbaues: Christoph und Unmack," *Tradition,* 14, 1969; the historical foreword to the catalog, *Nordischer Blockhäuser,* c. 1926; the various patent applications of both Doecker (sometimes written Döcker) and Christoph and Unmack; and various items in *Der Holzbau,* to which reference will be made.

40. According to German Patentschrift no. 13972, filed 7 October 1880 by Johann G. C. von Döcker, Copenhagen, and granted 17 June 1881. From 1880–84 patent applications were filed by Döcker not only in the major countries of Europe but also in the United States and in New Zealand.

41. According to German Patentschrift no. 26047, filed 29 April 1883 by Christoph and Unmack, Copenhagen, and granted 23 February 1884.

42. According to Wurm, "Vorgefertige Bauwerke des 19. Jahrhunderts," *Technikgeschichte Bd.,* 33, 1966, pp. 251–253. It would seem that Döcker's Hospital is that described in the *Scientific American,* June 1886, as Ducker's Portable Barrack and Field Hospital. Working drawings of this latter building are illustrated in Dietz, *Industrialized Building Systems in Housing,* 1971, p. 221.

43. "Prufung der von der Christoph & Unmack A.-G. Niesky O.-L. im Auftrag des Holzbau-Industriellen-Verbandes errichteten Wände (Block- und Tafelwand) auf Wärmeschutz," *Der Holzbau* (supplement to *Deutsche Bauzeitung*), 55:4, 1920, pp. 13–16.

44. "Holzhäuser der Siedlung Johannisthal bei Berlin, Ausgefuhrt von Christoph & Unmack," *Der Holzbau,* 54:1, 1920, p. 2; "Die Holzhaus-Siedlungen des "Wohnungs-Verbandes Gross-Berlin," *Der Holzbau,* 54:13, 1920, especially photo on p. 50 (Berlin-Steglitz) and p. 51 (Johannisthal).

45. "Bemerkenswerte Holzbauten von Christoph & Unmack in Niesky in der Ober-Lausitz," *Der Holzbau*, 54:19, 1920, pp. 75–76; also two illustrations of products of Christoph and Unmack in *Der Holzbau*, 54:22, 1920, p. 87.

46. Wurm, "Die Industrialisierung des Holzhausbaues," 1969, pp. 208–209.

47. Konrad Wachsmann, born 16 May 1901, in Frankfurt-On-Oder; died in Los Angeles, November 1980. Biographical details of Konrad Wachsmann are derived from curricula vitae prepared at various stages of his life and from the draft manuscript of his autobiography.

48. Specialized catalogs were produced for important building types. Two of these were *Krankenpavillons in Holzbauweise. Original-System Doecker*, Prospekt H.150, 1934; *Schulpavillons, System Doecker*, Prospekt H.161, 1933.

49. Blockhouse catalogs include: *Nordischer Blockhäuser*, Katalog XVII, c. 1926; *Nordischer Blockhäuser*, Katalog XVIII, 1934.

50. The Christoph and Unmack panel system is described in Dex Harrison, *A Survey of Prefabrication*, 1945, study sheet; and in Wachsmann, *Holzhausbau*, 1930, pp. 26–29.

51. As can be seen in the survey of the Kochenhofsiedlung, an estate of 22 all-wooden houses; Fritz Kress, "Die Konstruktionen und Holzbausysteme," *Moderne Bauformen*, 1933, pp. 596–634.

52. For example, Fritz Marcus, who designed a remarkably modern looking house on the Ruppenhorn, Berlin, which was published in *Moderne Bauformen*, 23, 1923, p. 231.

53. According to Wachsmann's autobiography and information transmitted by Mrs. Wachsmann.

54. Wachsmann, autobiography.

55. An illustration of a two-storied house undergoing this testing process appears in the article "Probe-Montage eines Holzhauses in der Montage-Halle von Christoph und Unmack," *Der Holzbau* (supplement to *Deutsche Bauzeitung*), 55:11, 1921, p. 42.

56. Even those built by Christoph and Unmack, such as the Einstein house (1929) near Potsdam, and the Children's Home (1928–29) in Spremberg. The system forms the basis of Wachmann's article: "Kleine und grosse Bauten in neuer Holzbautechnik," *Bauwelt*, 50, December 1931, pp. 1559–1574.

57. Wachsmann, autobiography.

58. Wachsmann, *Holzhausbau*, 1930, p. 7: "Maschinen in der Fabrik produzieren heute das Holzhaus, nicht der Handwerksbetrieb. Die alte hochentwickelte Handwerkskunst geht in die moderne Maschinentechnik über. Hier findet sie neue Möglichkeiten der Andwendung, neue Gestalten. Das Holz als Konstruktionselement in zimmermansmässiger Weise verarbeitet genügt nicht mehr allen Ansprüchen in fabrikatorischer und statischer Hinsicht. Aber als ein Konsequent von Maschinen bearbeitetes Fabrikationsmaterial hat es technisch und wirtschaftlich dieselbe Bedeutung wie irgendein anderer Baustoff."

59. Wachsmann's terms are: "Ortsfeste Fachwerkbauweise; Tafel- oder Plattenbauweise; Blockbauweise."

60. For these works see Wagner, *Das Wachsende Haus*, 1932.

61. For Bartning, see ibid., pp. 57–60; for Gascard and Canthal, ibid., pp. 112–115.

62. Spiegel, *Das Stahlhausbau*, 1928, pp. 142–154.

63. See the exhibition houses by Gellhorn, Mendelsohn, Von Steinbuchel-Rheinwall, Max and Bruno Taut, and Mebes, in Wagner, *Das Wachsende Haus*.

64. Reported in "Mitteilung aus der Fachwelt," *Moderne Bauformen*, 1929, pp. 72–78.

65. For this philosophy, see Wachsmann, *The Turning Point of Building*, 1961.

66.This was reported in the *Berliner Lokal-Anzeiger*, 18 March 1928. Sommerfeld had been the client for the famous wooden house built by Gropius in 1922. He was in the circle of the "Friends of the Bauhaus," which included architects such as Poelzig, and public figures such as Einstein. He had purchased the experimental Am Horn house from the Bauhaus and had also had "further business connections with the Bauhaus," according to Wingler, *Bauhaus*, 1969, p. 88. In 1927, just prior to this arrangement with Sommerfeld, Gropius had explored standardization in a series of wooden weekend houses.

67. The report in the *Berliner Lokal-Anzeiger*, 1928, referred in passing to Gropius' intention of studying building methods in America. In the Busch-Reisinger collection of Gropius drawings at Harvard University are two folders that contain the documents collected on this American study tour, showing with what conscientiousness he had undertaken the mission.

68. According to the minutes of a discussion between JFF and Walter Gropius, "Talk with Walter Gropius," typescript dated 8 June 1949, in the Bemis files.

69. There were many reports in the German press at this time of a visit made by Gropius, Sommerfeld, Wagner, city secretary Hirsch, and Ministerpräsident Braun to the Fichtalgrund housing area in 1928. The visitors talked to the press of the need for the development of standard housing types based on series production methods. Cuttings in the Gropius files include the *Berliner Lokal-Anzeiger*, 1928, referred to earlier; *Vassische Zeitung*, 29 March 1928; *Berliner Tagenblatt*, 19 March 1928, headed "Typisierung des Wohnungsbaues"; and *Karlsruhe Algemeine Zeitung*, 29 March 1928. An unidentified clipping shows a photo of the visitors and their families, formally posed.

70. Gropius drawings labeled "Montagehaus" (which in 1981 were in the process of classification in the Busch-Reisinger Museum).

Chapter Four

1. For an account of the historical development of the Hirsch copper company, see "Hirsch, Aron Siegmund (1858–1941?)," *Encyclopaedia Judaica*, 1972, v.8, and the note on Hirsch in "Metals and Mining," *Encyclopaedia Judaica*, v.11. The writer also in-

terviewed Mr. Sammy Oppenheimer, a former employee of Hirsch and exchanged letters with Mr. Siegmund Hirsch, then living in Lucerne, Switzerland. Information was also received from Mr. Gustave Hirsch, of Nahariya, Israel. Two books relate the story of the family and its business: Joseph Hirsch, *The Book of the Late Reb Binyamin Hirsch*, Jerusalem, Tarshish Publishers, 1948 (in Hebrew); Siegmund Hirsch, *Revolution in Messing: 1908–1928*, Lammersdorf, Otto Junker, 1967. The novel, *Shaul and Yohanna*, by Naomi Frankel (in Hebrew) gives a fictional, but essentially realistic, picture of the family and its times.

2. Walter Gropius, *Internationale Architektur*, 1925 (reprinted in *Images*, Open University, 1975).

3. Hirsch exhibited at the Bauwelt-Musterschau in Berlin in 1932, as part of the group "Dacher und Richteifen-Metalle," together with the German Copper Institute, the Luxfer Prism Co., and other manufacturers of building materials and components. (See *Bauwelt*, 42, 20 October 1932, p. 1063).

4. Förster was probably an engineer; Krafft may have been the architect Robert Krafft, who won an award in the Berlin Wachsende Haus competition of 1931–32.

5. In 1924, in Budapest, Frigyes Förster had filed an application for his invention for wooden-framed metal-clad panels, fitted with insulation of a light-weight material, and this had been granted as an Austrian patent (112926) with effect from 15 December 1928. There are of course precedents for metal-faced, heat-insulating walls even before this, perhaps originating with refrigeration technology. For instance, James C. Woodson, on behalf of the Westinghouse Corporation, filed an application for an insulating sandwich wall which in some ways predates even the first of the Förster patents (U.S. 1,626,655). Förster acknowledges the precedents set by the designers of storage rooms and refrigeration plants. Our account of the patent given here derives from the English version of the American patent, filed 25 April 1925, and granted 11 September 1928 (1,683,966).

6. The revised system was submitted by Friedrich Förster and Robert Krafft to the German patent office on 25 August 1930 and was eventually granted on 14 April 1932, retroactive to 26 August 1930 (DRP 548532). A supplementary patent was granted 5 April 1934, retroactive to 24 December 1930 (DRP 595292). The invention was also granted patents in England (357356), America (1,965,636), France (704417), Belgium (374,558), Canada (326,458), Hungary (104485), Holland (30749), Norway (51242), Ireland (11756), Poland (14107), and Australia (25763/30). The early precedent to the Förster and Krafft system is described in "Holzhäuser der 'Holzbau A.G.' in Neuss am Rhein," *Der Holzbau* (supplement to *Deutsche Bauzeitung*), 56:2, 1922, pp. 5–8, and is illustrated on p. 8.

7. From the published descriptions, it would appear that this house was based on the original, rather than the revised, Förster system. See "Portable Steel and Copper House Developed in Germany, *The Iron*

Age, 12 February 1931, and "Germany Now Has Copper Houses Made in Sections Easily Set Up," *Copper and Brass Bulletin*, c. 1931.

8. A note in the Hirsch catalog, *Kupferhaus: Das Ideale Einfamilienhaus*, 1931, states: "Ausgezeichnet mit der hochsten Auszeichnung auf der Internationalen-Kolonialaussteilung Paris 1931, dem 'Grand Prix,' "

The Hirsch copper house is described in *Bauwelt*, 20, 14 May 1931, p. 672, and illustrated and described in *Deutsche Bauzeitung*, 15 July 1931, p. 347. The Berlin building exhibition, at which the copper house was shown, included the exhibit "Die Wohnung Unserer Zeit," which was a showcase for the Modern Movement, with Mies van der Rohe's famous exhibition house. For this exhibition, see *Bauwelt*, 9, 26 February 1931, pp. 14, 28, 42; *Bauwelt*, 19, 7 May 1931, p. 664; *Deutsche Bauzeitung*, 6 May 1931, pp. 217ff.; and " 'Die Wohnung Unserer Zeit' auf der Deutschen Bauaustellung Berlin 1931," *Bauformen*, July 1931, p. 329ff.

9. Letter from Rudolf Weilbier, editor of *Bauwelt*, to Grünfeld, of the Deutsches Kupfer-Institute (30 May 1931).

10. Letters Joseph Wormser/Gropius (20 May 1931); Gropius/Wormser (22 May 1931). Gropius deals with other criticism in various letters. To Siegfried Hirsch (4 July 1931) he writes of the criticism of Hegeman, and of a critical article in *Industriekurier*; in a letter from Grünfeld to Gropius (4 December 1931), the director of the Copper Institute quotes a sharply critical letter from a Herr Metzger, which he is sure Gropius could rebut.

11. *Bauwelt*/Gropius (12 June 1931); Gropius/*Bauwelt* (16 June 1931).

12. There is a possibility that Gropius had formal connections with Hirsch Kupfer even prior to June 1931. We find him checking the accounts for building work in-situ carried out for the Berlin building exhibition (see letter Gropius/Hirsch Kupfer 18 June 1931)—and we know that the exhibition opened in May.

13. Gropius/Schawinsky (11 July 1931): "Ich habe neuerdings die Bedreuung der Kupfer-Häuser der Hirsch, Kupfer-und Messingwerke A. G. Übernommen." Gropius/Frank'sche Eisenwerke (12 August 1931): "Ich habe vor einiger Zeit die Leitung des neuer Kupferhausbaues der Hirsch, Kupfer-u. Messingwerke . . . Übernommen und arbeite diese Materie technisch und künsterlisch volkommen neu durch."

Gropius/Ministry of Posts (15 August 1931), considers himself "leiter die Abteilung Kupferhausbau" of Hirsch Kupfer. He uses the identical phrase in a letter to Max Knapp (28 August 1931).

14. "Gutachten über die Kupferhäuser der Hirsch, Kupfer- und Messingwerke A.G., Charlottenburg 2, Hardenbergstr. 43" (19 June 1931).

15. This is supplemented by further comments in a "Nachtrag" of 15 July 1931, after further careful observation of the copper houses at the building exhibition.

16. Organizations and institutions consulted in the process of testing the copper houses included: Materialsprüfungsamt, Dahlem (fire safety); Preuss. Landesanstalt für

Wasser-, Boden- und Lufthygiene (problems of patina); Spezialsachverständiger Saemann, Bremer; and Collignon, Berlin (lightning protection); Heinrich-Hertz-Institüt für Schwingungsforschung, Berlin (acoustics); Forschungsheim für Warmeschutz, Munich (condensation and heat control). Constant contact on technical matters was also maintained with the Deutsches Kupfer-Institut, Berlin.

17. Among the stoves proposed by Gropius were the Oranier anthracite models, which he himself had designed, and which were among the Bauhaus' most successful earners of royalties.

18. In this respect see the interchange of correspondence between Gropius and W. Kurz of the Aluminium-Walzwerke at Singen, 3 to 5 March 1932.

19. Gropius was in contact with Böhler-Stahlbau (12 August 1931); with Vogel and Noot, of Steiermark (11 February 1932); with von Halem Stahlwerksverband, of Düsseldorf (12 February 1932) and Schneider of the Technisches Büro für Stahllamellenbau (22 February 1932).

20. Gropius advocated a l-m module, which was a useful dimension, suiting both doors and windows (handwritten office memo, 6 October 1931). He returned to the problem of modularity in working notes of 9 January 1932.

21. For the Weissenhofsiedlung at Stuttgart, Gropius had built a house using a lightweight steel frame, clad with asbestos, and filled with cork insulation. Now he wrote to Hirsch (4 July 1931): "The material is without doubt good for heat insulation, but not as good as metal foil."

22. This was presumably extended for a further year, being finally cancelled on 31 December 1932 (letter Hirsch Kupfer/Gropius, 20 June 1932).

23. Gropius/Hirsch Kupfer, 4 July 1931.

24. The consolidated price list, excluding foundations, freight and erection charges, included all the types A through H, J, K, and R, as at 21 July 1931.

25. These dimensions are net and are taken from the plans. They differ slightly from areas given in various schedules. Areas also vary from one sketch plan to another, but the differences are not significant.

26. All types are referred to in various documents, dated from 18 to 30 July 1931. At one stage there is reference to a type L, but later, on 21 July, this is altered by hand to type K. There are L type drawings, but at least on one there is an indication of change to type K.

27. On 18 August, a revised price list for types K and M was issued, including the cost of models with an optional basement.

28. All drawings, and many prints, are in the Gropius collection in the Busch-Reisinger Museum.

29. Gropius/Hirsch Kupfer, 4 July 1931: "When ten houses per month are built, could we save workers?"

30. "Aktennotiz zur Besprechung am 21.10.31 in Eberswalde."

31. "Entwurf eines Arbeitsplanes für die zum frühjahrsverkauf fertigzustellenden typen lt. bauprogramm 1932 Typ 1–5", ("Aktennotiz Hirsch-Kupfer," dated 8.1.32.).

32. Gropius' activities on behalf of Hirsch Kupfer were not limited to the copper houses. As he wrote to his friend Alexander Schawinsky (11 July 1931): "The sales manager Wallach yesterday asked me, at Eberswalde, to give him ideas for the sale of other products of Hirsch Kupfer . . . products such as copper profiles, sheets, and piping."

33. These sources included a list of customers, prepared in mid-July 1931, giving names, house types, locations, time of delivery, sums deposited, and status of the project; a note by Gropius of a visit to two houses under construction, at Schildow and Tegel Konradshohe, on 23 October 1931; and letters of appreciation in the Deutsche Kupferhaus Gesellschaft catalog, many of which can be positively identified on the lists, leading to the supposition that they are actually Hirsch houses.

34. Hirsch Kupfer/Gropius, 15 July 1931.

35. According to a discussion between Gropius and JFF of the Bemis Foundation, recorded on 8 June 1949, Gropius stated that he had worked for Hirsch Kupfer "and they erected 53 houses."

36. Letter Gropius/Mandl, 30 April 1932.

37. Letter Thiess & Co./Hirsch Kupfer, 15 July 1931.

38. Letter Gropius/Hirsch Kupfer, 13 July 1931.

39. Williams is referred to in a letter to Hirsch Kupfer on 9 July 1931, and both Williams and Hoboken appear in the list of customers of mid-July.

40. According to a note on the reorganization of Hirsch Kupfer in *Berliner Tageblatt* of 27 April 1932.

41. The Hirsch letter dealing with the Katanga interest is dated 15 August 1931. There is no date on the translated memorandum "Einige Beobachtungen bezüglich all-Kupferhäuser" which emanated from the Department Maisons Tout Cuivre of Union Miniere du Haut-Katanga of Brussels.

42. The Anaconda Company is referred to in the July list of customers, so presumably contact was made in June 1931 or earlier. A long, detailed letter was sent on 17 August 1931, referring to previous cables and other correspondence, and this was translated and kept in Gropius' files.

43. There had already been detailed discussions with the Ostdeutsche Eilschiffart G.m.b.H., for a trial shipment of three copper houses to America, via the port of Hamburg, by 1 August 1931. In an undated (presumably subsequent) document there is a detailed breakdown of weights and measures, both for shipping and customs purposes.

44. Letter Gropius/Max Knapp, Union Guarantee & Industrial Trust Ltd., Budapest, 28 August 1931; reply 31 August 1931; subsequent letter 7 September 1931.

45. Letter Gropius/Hirsch Kupfer, 2 August 1931.

46. C. Borngräber, "Foreign Architects in the USSR," *Archi-*

tectural Association Quarterly, 11:1, 1979, p. 52, notes that "at the beginning of the 30s, 90 per cent of the architects were unemployed. 1400 architects applied in Frankfurt for a transfer to Moscow."

47. Letter Gropius/Ernst May, 6 August 1931

48. The four-page, unsigned, memorandum dated 6 July 1931, has many corrections in pencil, and has the names HKM (Hirsch Kupfer-und Messingwerke) and Malletke written in the margin.

49. In a letter to Wallach, sales manager of Hirsch Kupfer, on 24 July 1931.

50. Letter from Schloss, of Hirsch Kupfer, to Moufang, dated 1 February 1932.

51. Letter Hirsch Kupfer/Moufang, 3 March 1932.

52. "Verwendung grössmöglicher Typiesierung und grösstmöglicher Variation, in dem nach Analogie eines Baukastens im grossen die Bau*teile* genormt oder zu verschiedenen Hauskörpen den jeweiligen Bedürfnissen entsprechend zusammengesetzt werden können," Gropius, in draft of letter to Dr. Schmidt of the Ministry of Labour, 20 June 1931.

53. A series of articles in *Bauwelt* during October 1931 point to the growing interest in core houses ans staged growth. M. Schadewald, in "Wachsendes Haus—Sinkende Schuld," 41, 8 October 1931, pp. 1291–1293, gives a history of *Staffelhäusern* from 1925 to 1927, and illustrates a model house, showing staged development; Dirk Gascard and P. M. Canthal, in "Das Kleinst-Eigenheim," 42, 15 October 1931, pp. 1324–1325, deal with the concept of the core house; architect Murche, of Berlin, in "Doppelhaus als 'wachsendes' Kleinsiedlungshaus," 44, 29 October 1931, deals with the expanding two-family house, and in the same issue Ella Briggs, "Praktische Fragen zur Erwerbslosensiedlung," pp. 1394–1396, deals, inter alia, with the growing house concept, and in "Kleinsthäuser," pp. 1397, we have further discussion of the core houses of Schadewald and Gascard. In Wagner's book *Das Wachsende Haus* a core house by architect Schutz is illustrated which dates back to 1921.

54. In October 1931, "the government allotted some 78 million Marks for the construction of 'Stadtrandsiedlungen' " . . . as part of a full-scale 'back-to-the-land' movement." Anderson, *Architecture as a Means for Social Change in Germany, 1918–33*, 1972, p. 116; see also p. 117.

55. The competition was extensively covered in the journals of the day. The announcement first appeared in *Bauwelt*, 43, 22 October 1931, p. 1376, and details of the proposed exhibition appeared in *Bauwelt*, 49, 3 December 1931, p. 1554. The results were published, together with a criticism of the prize-winning schemes, in *Bauwelt*, 1, 7 January 1932, pp. 3–7. The winning schemes were illustrated in *Bauwelt*, 2, 14 January 1932, and in *Deutsche Bauzeitung*, 13 January 1932, pp. 55–60. Details of the task set, the *Arbeitsgemeinschaft*, were given in *Bauwelt*, 1, 7 January 1932, pp. 3–4; and a detailed discussion, together with illustrations of the work, appears

in Martin Wagner, " 'Das Wachsende Haus' der Arbeitsgemeinschaft," *Deutsche Bauzeitung*, 13 January 1932, pp. 41–43, 53–55.

56. Christian Borngraber, "The Socialist Impact of the New Architecture in Germany," *Architectural Association Quarterly*, 11:1, 1979, p. 42.

57. There were 23 houses, according to the announcement in *Bauwelt*, 17, 28 April 1932; an account of the opening of the exhibition in *Deutsche Bauzeitung*, 11 May 1932, p. 381, however, states that there were 30 built and fully equipped examples of the Growing House, and 79 models and designs from the competition. In Wagner's book (note 59), there were 27 houses by 23 architects, whereas a siteplan in Gropius' possession showed 26 houses by 22 architects. The lists are similar, but not identical.

58. In *Bauwelt*: "Haus für alle?", 22, 2 June 1932, pp. 539–540; "Technische Einzelheiten auf der Berliner Sommerschau," 24, 16 June 1932, pp. 590–591; "Ein Baufachman geht durch die Sommerschau," 26, 30 June 1932, pp. 635–636; "Das Wachsende Haus," 28, 14 July 1932, p. 688. In *Deutsche Bauzeitung*: Erich Petzold, "Zum Problem des Wachsenden Häusen," 17 February 1932, p. 141; Erich Heinecke, "Das 'Wachsende Haus,' " 11 May 1932, pp. 392–398; Heinecke, "Zum Thema 'Wachsende Haus,' " 22 June 1932, pp. 503–510; Hans Spiegel, "Stahlbauweisen für das kleine Einzelwohnhaus," 6 July 1932, pp. 553–560; W. Ludowici, "Die Finanziellen Vorteile des Wachsendes Hauses," 20 July 1932, p. 584.

59. Martin Wagner, *Das Wachsende Haus*, Berlin, Deutsches Verlagshaus Bong & Co., 1932.

60. The Vienna Growing House Competition was held at the beginning of 1932, the results being announced on 18 February 1932. A comprehensive illustrated report followed: Max Eisler, "Das 'Wachsende Haus' in Wien," *Moderne Bauformen*, 1932, pp. 289–308.

61. Gellhorn adapted the system for dry construction, calling it "reine Trockenmontage."

62. For an account of this construction see Spiegel, *Das Stahlhausbau*, 1928, pp. 121–124.

63. This was probably the Dyckerhoff and Widmann system, also used by Mendelsohn as an alternative proposal to his Böhler house. We have already noted Dyckerhoff's role in the quest for a better insulation system for the Hirsch houses.

64. Martin Wagner, *Prefabricated Housing*, 1941, pp. 71–72.

65. Letter Gropius/Kurz, Aluminium-Walzwerke, Singen, 5 March 1932.

66. The task-force houses were first published in January 1932, when sketch plans were shown in *Deutsche Bauzeitung*, 13.1.32, pp. 41–55.

67. There is one L-shaped plan for a five and a half roomed house, embracing a patio, among the Gropius sketches for copper houses. It is drawing no. 25, unidentified and with no date, in the collection in the Busch-Reisinger Museum.

68. The drawings (166 to 186) all have the Hirsch Kupfer stamp. The basic set—plans, sections, elevations—are undated. The details date from 7 to 19 March 1932.

69. Letter Hirsch Kupfer/Gropius, 21 April 1932.

70. Letter Hirsch Kupfer/Gropius, 19 May 1932.

71. For instance, in the *Architectural Forum*, which as early as 6 February 1932 was cabling Gropius, pressing him for material. It was eventually published under the heading "Portable Houses in Copper," in February 1933 (pp. 145–147).

72. "Nach einem Patent von Förster und Krafft von der Hirsch Kupfer-und Messingwerke," in Gropius' text in Wagner, *Das Wachsende Haus*, 1932, p. 65.

73. Walter Gropius. "The Formal and Technical Problems of Modern Architecture and Planning," *The Architect and Building News*, 18 May 1934, p. 205.

74. "Die von mir erstmalig in einer Broschure uber die Industrialisierung des Hausbaues in Jahre 1910 ausgesprochene, inzwischen vielfach befehdete Idee, Häuser in stationären Werkstätten in Ihren Teilen serienmässig herzustellen und daraus variable Typen wie aus einem Baukasten im grossen zusammenzusetzen, wird heute endlich der Verwirklichung entgegengeführt." Gropius, in Wagner, *Das Wachsende Haus*, 1932, p. 65.

75. Ibid.

76. The accounts submitted on 27 May 1932 totaled RM 7,598. Of this sum the two major items were the cost of producing houses in the factory (materials, salaries, expenses, transportation, assembly), RM 4,260, and on-site work (including bricklaying and carpentry), RM 2,269. The latter work was undertaken by the Allgemeine Häuserbau Company of Adolf Sommerfeld, for whom Gropius had once built a widely publicized wooden house and with whom he had been associated in a housing company.

77. See Mechtild Strattman, "Housing Policies in the Weimar Republic," 1979.

78. Ibid., pp. 21–22.

79. This inquiry was reported under the headline "Hirsch Kupfer. Reorganisation unter Auslandführung," in *Berlin Tageblatt*, 199, 27 April 1932.

80. Letter Hirsch Kupfer/Gropius, 2 May 1932.

81. "Nicht Mehr Kupferhäuser," *Bauwelt*, 30, 28 July 1932, p. 744.

82. Letter Gropius/Lauber, 27 April 1932.

83. Letter Gropius/Mandl, 3 May 1932.

84. See the letter from Mandl to Gropius, of 12 June 1932, for details of these negotiations.

85. Between a half to one million Reichsmarks for the rights to produce, but not the patents of Förster and Krafft.

86. Letter, Dr. Pleuss of the legal department of Hirsch/Gropius, 20 June 1932.

87. This correspondence, over the period 3 December 1932 to 20 June 1933, deals with the sale of the Hirsch houses in the Berlin exhibition and with

requests to Gropius to supply the necessary documentation for their re-erection elsewhere. No further information on the Berlin Ilsenberg Metallwerke is revealed by the Gropius/Hirsch files, but it may be relevant to note that one of the Hirsch copper works was located at Ilsenberg.

88. According to Joseph Hirsch, *The Book of the Late Reb Binyamin Hirsch*, 1948.

89. The Germans found it difficult to dispense with the expertise of many of the leaders of the industry and appointed, on 10 July 1933, an Advisory Committee to the Metal Industry of six members (all of whom were Jewish) which included Mendl Hoffman, as representative of Erze und Metalle Hirsch A. G. (*Palestine Post*, 25 July 1933, p. 5).

90. For the relationship between Hirsch and AEG, I am grateful to Ms. Eva Wirtz, of AEG-Telefunken Central Archives, for the information supplied from the firm's records. An account of the early connections between the two firms is given in Siegmund Hirsch, *Revolution in Messing*, 1967, especially the chapter "Siemens und Allgemeine Elektricitätsgesellschaft (AEG)," pp. 178–183.

91. The establishment of "Hirsch Kupfer-und Messingwerke G.m.b.H., of Frankfurt, in 1947, is recorded in "Das Spezial-Archiv der Deutschen Wirtschaft" (Auszug a.d. firmenkundlichen Bericht, 1967).

92. René Schwartz was the son-in-law of Aron Hirsch and manager of the subsidiary company, Hirsch Revisions G.m.b.H. In a letter to the writer on 16 March 1978, Mr. Siegmund Hirsch indicated that his late brother-in-law, Mr. Schwartz, had played a role in the copper house project of Hirsch Kupfer and initiated the later export of the houses to Palestine.

93. Letter Gropius/Schwartz, 24 November 1932.

94. This circular letter was sent to Gropius on 2 December 1932 by Dr. Luttke of the "Gesamtausshuss zur Wahrung der Interessen der deutschen Metallwirtschaft." It was accompanied by pages from the Deutsche Kupferhaus catalog.

95. As reported by Albrecht Hesse to Walter Gropius in a letter of 12 December 1932.

96. Gropius expressed his skepticism in a private note to the editor of *Neue Linie* of 15 December 1932. Regrettably, he wrote, his own houses were now not available. To Schwartz, on the same day, he wrote in controlled anger, protesting the belittling of his work. "The types that have been developed by me," he pointed out, "have been examined much more profoundly and are further from the experimental stage than those types which I found when I started to work with Hirsch Kupfer."

Chapter Five

1. According to the *Post*, 1,200 immigrants were expected. Between March and September 1933, 6,000 German Jews settled in Palestine. (*Palestine Post*, 27 Sept. 1933, p. 2.)

2. In January 1933, 2,371 immigrants entered Palestine, of whom 2,249 were Jews. Seventy-three entered with £P 1,000, and a further 462, who had "less assets," were nevertheless admitted (*Pales-*

tine Post, 10 March 1933). According to the Draft Immigration Ordinance, published in the *Palestine Gazette*, of July 1932, "Category A" comprised persons of independent means, that is, possessing £P 1,000 and over, or a person following a liberal profession who had £P 500 or more, or a craftsman with £P 250 or more, if according to the British authorities, there was a need for his skills in Palestine (according to the *Palestine Post*, 4 August 1933, p. 1).

3. This account of the transfer agreements is derived from "Haavara," *Encyclopaedia Judaica*, v.7, pp. 1012–1013; Werner Feilchenfeld, *Jewish Trade Policy: On the Basis of Transfer Agreements with Central and East European Countries*, Tel Aviv, Haaretz Press, 1938; David Yisraeli, "The Third Reich and the Transfer Agreement," *Journal of Contemporary History*, 2, 1971, pp. 129–148; items in the *Palestine Post*, 7 September, 28 December 1933; Ernst Marcus, "The German Foreign Office and the Palestine Question in the Period 1933–1939," *Yad Vashem Studies*, 2, 1958, pp. 179–204.

4. Yisraeli, "The Third Reich," 1971, p. 129.

5. Yisraeli, "The Third Reich," 1971, p. 130.

6. "Haavara," *Encyclopaedia Judaica*, v.7, pp. 1012–1013.

7. Yisraeli, "The Third Reich," 1971, p. 130.

8. For an excellent account of German attitudes toward the transfer agreements, see Yisraeli, "The Third Reich," 1971; Marcus, "The German Foreign Office," 1958, gives a personal insight into German policies at the time.

9. The transfer agreement, an "unholy alliance," in the words of one commentator (*Palestine Post*, 20 September 1933), was severely criticized at the Zionist Congress in Prague, in 1933 (according to *Palestine Post*, 28 December 1933, p. 7); however, the "controversy was settled at the Zionist Congress in Lucerne (1935) which decided by a vast majority in favor of the transfer . . ." (*Encyclopaedia Judaica*, v.7, pp. 1012–1013).

10. "Haavara," *Encyclopaedia Judaica*, v.7, p. 1012–1013.

11. Feilchenfeld, *Jewish Trade Policy*, 1938, p. 14.

12. *Palestine Post*, 7 September 1933.

13. Feilchenfeld, *Jewish Trade Policy*, 1938, p. 16.

14. Gilbert Herbert, *Pioneers*, 1978.

15. J. Shiffman, "The Building Industry in Palestine," *Palestine and Middle East Economic Magazine*, 7–8, 1933, p. 288.

16. Many references are made to the boom and its consequences in the *Palestine Post*; a review of the situation is given by Shiffman, "The Building Industry," 1933, p. 288.

17. "An enormous increase in the demand for building materials was again noticeable in October. . . . Imports for wood and timber, for instance, rose from £P 14,800 to £P 25,000, of cement from £P 2,000 to £P 10,300, and of iron bars and girders from £P 7,300 to

£P 32,400'' (*Palestine Post*, 4 March 1934, p. 3).

18. In the *Judische Rundschau* of 11 August 1933, an article on land and building costs in Haifa brought home the problems of conventional building in Palestine—facts that were probably read with interest both by the manufacturers of prefabricated houses and by their potential customers.

19. *Judische Rundschau*, 30 June 1933, p. 296.

20. The first advertisement for the Deutsche Kupferhaus Gesellschaft appeared in *Judische Rundschau* on 30 June 1933. Advertisements continued fairly regularly, reaching a climax in December when, at weekly intervals, an impressive illustrated advertisement appeared—the last on 29 December 1933.

21. The following announcement was published in the Palestine catalog: ''Nach der Verfugung des Herrn Reichswirtschaftministers, Abtig. Devisenbewirtschaftung an unsere Gesellschaft vom 24 Juli 1933, Akt. Z. Dev. 1.28857.33, ist auswanderern nach Palastina der Erwerb eines deutschen Kupferhauses und die Mietnahme als Umsugsgut gestattet, ohne das eine Anrechnung auf die als Vorzeigegeld von der Einwanderungsbehorde verlangten Pal. L.1000 erfolgt.''

22. The Deutsche Kupferhaus Gesellschaft advised readers of the *Judische Rundschau*, in their advertisement of 4 August 1933, that ''Unser Palästina-Katalog erscheint am 10 August 1933.''

23. The first mention of an agent appears in *Judische Rundschau* on 5 September 1933 and his name, Fritz Neumann, is given on 8 December 1933.

24. I am indebted to Mrs. N. Markowicz for information about her late husband's connection with the copper houses (interview 10 May 1977).

25. For an expression of this concern, see the following correspondence: Dr. Georg Landauer/Palastine-ampt., Berlin (2 January 1934); Palästina-ampt./Jewish Agency, Jerusalem (11 January 1934). (Jewish Agency Archives, File S7/83.)

26. A certain Felix Hapner (Hupner? Hymer?—the name is unclear) had proposed to the Jewish Agency that copper houses be exported and erected there at cost (no profit being sought), on land to be provided by the Jewish National Fund. After consideration, the agency found it advisable to turn down this proposal. See letters Hapner/Dr. Ruppin (14 December 1933); Landauer/Hapner (13 March 1934). (Jewish Agency Achives, File S7/83.)

27. These were motivated by a confidential report from London, quoting doubts as to the suitability of the copper houses for Palestine, expressed by a man who had previously held a leading position with Hirsch Kupfer, and by requests for information on the Haifa experience, because ''many emigrants have invested money in copper houses.'' This information is contained in the following correspondence: Landauer/Palastina-ampt. (21 January 1934); Palastina-ampt./Jewish Agency, Jerusalem (11 January 1934). (Jewish Agency Archives, File S7/83.)

28. See letter Landauer/Hitachdut Olej Germania, Haifa, (22 January 1934). (Jewish Agency Archives, File S7/83.) It is not clear whether the cost analysis was initiated by the agency or the Hitachdut. After assessing all relevant costs of the "Haifa" model, including transportation, erection charges, and the initial investment, the report concluded that the copper house was expensive and economically not worthwhile. Independent sources (such as Mrs. Markowicz) confirm the high cost of erection.

29. The Tuchler house in Haifa, for instance, was granted a temporary permit in April 1934. Each year, an extension of this permit was applied for, until June 1939, when the City Engineer finally granted the copper house permanent status.

30. Letter Dr. Senator/Palastina-ampt. (5 August 1934). Jewish Agency Archives, File S7/83.)

31. According to the account of Mr. Ruckenstein, whose father imported one of the houses, there were eleven houses in Haifa, one in Safed, one in Ramat Gan, and one at Petah Tikvah which was not erected but sold immediately because of the high value of the copper material. The Hitachdut Olej Germania (in their letter of 22 January 1934) refer to seven or eight houses in Haifa.

32. Information about the two houses at Ramat Gan was obtained from Mr. Uri Schiller, architect, formerly of the Ramat Gan Municipality.

33. According to the Minute Sheets of Haifa Municipality (Folio 1,4996/34, March 1934) special permission had been granted for six copper houses to be erected in Haifa:

Schoenfeldt house, 5 Leonardo da Vinci St., Haifa (Jaffa type). Permit 21.2.34.

Grundmann house, 9 Hurcha St., Haifa (Libanon type). Permit 22.2.34.

Neumann house, 147 Sea Rd., Haifa (Libanon type). Permit 10.4.34.

Tuchler house, 20 Tel Mane St., Haifa (Jerusalem type). Permit 26.4.34.

Kaliski house, 3 Tamar St., Haifa (Haifa type). Permit 14.5.34.

Shmukler house, site unknown, Haifa (dismantled c. 1972).

In addition, the following house was erected in Safed:

Ruckenstein house, Mt. Canaan, Safed (Haifa type), c. 1934.

34. According to information supplied by Mr. Eichbaum, of Haifa.

35. The original copper roof of the Kaliski house corroded, because iron instead of copper nails were used in its construction, and was eventually replaced with a tiled roof.

36. "Stahl-beton haus, transportabel, für Palästina besteeignet, kein Blech, massiv, fungenlos bewart. la Fabrikat. Als umzugsgut in Deutschland zahlbar." (*Judische Rundschau,* 5 January 1934, p. 10.)

37. History repeats itself: the term "portable" was used to describe the first wooden prefabs (the "portable colonial cottages") sent out in the 1830s from Britain to Australia (see the account in Herbert, *Pioneers,* 1978, ch. 2).

38. *Judische Rundschau*, 22 December 1933, p. 1012.

39. Margarete Sallis-Freudenthal, in her autobiography, *Ich Habe Mein Land Gefunden*, Frankfurt-am-Main, Knecht, 1977, p. 163, tells of the bringing of a portable timber house to Palestine from Germany.

40. House for Menahem (Max) Naumburg, 24 Smolenskin Street, Haifa. Information comes from the municipal files and from documents kindly supplied by Mr. Naumburg. Discussions were also held with Mr. Eichenbaum, who had worked upon the assembly of this house.

41. Naumburg had seen the special supplement of the *Frankfurter Zeitung* of 9 May 1935, "Neuzeitlicher Holzhausbau," which not only had an advertisement of Christoph and Unmack but mentioned the firm editorially.

42. Letter Rath/City Engineer, Haifa, 10 March 1938.

43. Drawings with the elements numbered to ensure proper sequence of erection were supplied by Christoph and Unmack, and the whole process went smoothly, except that, according to the owner, the waterproofing was applied incorrectly and subsequently had to be replaced with galvanized corrugated iron. The balconies, according to Eichenbaum, needed strengthening with iron some years ago.

44. *Judische Rundschau*, 15 August 1933, p. 431. See also the letter from architect Curt Leschnitzer, Berlin, in *Judische Rundschau*, 8 September 1933.

45. *Judische Rundschau*, 15 September 1933, p. 542.

46. The Kletzin steel house, according to the *Berlin Lokalanzeiger* of 25 March 1934, was made of steel sandwich panels well insulated, and finished in white or aluminium color.

47. See the report "Sectional Houses Developed in Germany," *Architectural Record*, July 1936, p. 70.

48. See, in the Jewish Agency Archives (Jerusalem) File S7/83, the following items: Georg Breslauer, description of a metal house, 6 March 1934; letters Palastina-ampt./Jewish Agency Jerusalem and Dr. Senator (21 March 1934, 23 July 1934) re Grundrisslose Metalhäuser; and letter Dr. Senator/Palastina-ampt., on same subject, 5 August 1934.

49. Breslauer's memorandum is on file in the Jewish Agency Archives, but the accompanying photographs and drawings appear to be missing.

50. There is a house with a prefabricated steel frame (of a type different from that of Böhler but not yet identified) which was erected in 1937–38 at 9 Witkin St., Haifa, and another (drawn to my attention by Haim Shapiro) in Ramat Gan.

51. The catalog *Ein Haus für Sie in Palästina* was published by the Export Bau-und Handels-Gesellschaft on 2.12.34. in Berlin. In addition, in 1935, the Palestine Building Syndicate put out a mimeographed specification and price list, breaking the costs into German and Palestine components.

52. I am indebted to architect Dov Eitan (formerly Entin) of Haifa, for valuable information

about the Böhler system in Haifa. His recollections supplement the contemporary account of the building of the steel structures given by his partner Kalitzky in the Palestine building journal, *Habinyan Hamisrach Hakarov*, 2.1, 1937, pp. 10–11. The two examples so far traced in Haifa are

Heimann Building, 39 Aviv (now Kadimah) St., Haifa. Permit 14.10.1935. Architects: Kalitzki and Entin.

Levin Building, 43 Hillel St., Haifa. Permit 21.1.1937. Architects: Kalitzki and Entin.

53. *Bauwelt*, 46, 17 November 1932, p. 1160.

54. I am grateful to Mrs. Max Loeb for allowing me access to her late husband's files. The following correspondence relates to the Böhler system: letters from Weltsch and Heinemann, 24.9.35, dealing with the design of five apartment blocks for Frau Levy-Stern; the letter of Dr. Toni Stern, Dortmund, c. December 1935, to Heinemann. Of particular importance is a comprehensive set of drawings of apartment buildings with a modular structure and based on Böhler prototypes in the Palestine catalog, jointly signed by Loeb and Heinemann.

55. My information on the Böhler system houses in Jerusalem comes from my discussions with Dr. Walter Katz, owner and long-time resident of one of the structures; Dr. Fritz Schindler, formerly a director of the Palestine Building Syndicate Ltd.; and Mr. Karl Rosenthal, formerly Jerusalem manager of the same company. The three houses identified by Mr. Rosenthal in Jerusalem, all designed by architect Dov Kuzinsky, are Katz house, Rehavia; Schiller house, Talpiot; and Freund house, Bet Hakerem. According to architect Dov Eitan, of Haifa, some Böhler houses in Jerusalem were designed by architect Heinz Rau, but these have not been traced.

56. Katz house, 9 Palmach St., Rehavia. Municipal plans signed by owner, 13.1.1937, approved by municipality 21.2.1937; working drawings by architect Kuzinsky, stamped "Palestine Building Syndicate," variously dated 2.1.1937, 20.1.1937, 1.2.1937; miscellaneous plans annotated in German, stamped "Export Bau-und Handelsges," dated 2.2.1937.

57. "The rate of exchange was adjusted from time to time by the Haavara according to the disagio, necessitated by the subsidy which the Haavara granted the Palestinian importers, to make up for the steadily deteriorating value of the Reich mark, so the German goods could compete with other imports. The ensuing disagio, borne by the emigrants, accordingly increased from 6% in 1934 to 50% in 1938." ("Haavara," *Encyclopaedia Judaica*, v. 7, pp. 1012–1013).

58. According to Ester Barsky-Krauss (in a report to the author), these buildings, since demolished though in excellent condition, stood at 16 Bialik Street and 4 Sprinzak Street, Holon.

59. Throughout the month of June 1934, an advertiser in the *Palestine Post* offered for sale "a splendid 10-roomed (insulated) copper house, which can be erected anywhere ("cool in summer, warm in winter, can be erected within two months"—an expanded version of the advertisement reads) to be sold under the

most favorable terms." These announcements, which were published in the German language, presumably refer to a copper house already imported into Palestine but not yet erected. The advertisements appeared at least five times, indicating that the owner was experiencing some difficulty in disposing of his property.

60. Letter Gropius/Von Lacsay, patent attorney, 21 February 1934, in which he discusses "the difficult situation of the Deutsche Kupferhaus Gesellschaft."

61. Feilchenfeld, *Jewish Trade Policy*, 1938, p. 16.

Chapter Six

1. Sibyl Moholy-Nagy, "The Diaspora," *Journal of the Society of Architectural Historians*, 24:1, 1965, p. 24.

2. According to the 15-page protocol of the infamous Wannsee conference, reported in Ya'acov Friedler, "Remembering Wannsee," *The Jerusalem Post*, 19 January 1982, p. 5.

3. Cecil Roth, *History of the Jews*, New York, 1961, p. 382.

4. Moholy-Nagy, "The Diaspora," 1965, p. 24.

5. Barbara Miller Lane, *Architecture and Politics in Germany, 1918–45*, Cambridge, 1968, p. 180. The relationship of the National Socialists to the Modern Movement in architecture in Germany, with all its complexities and contradictions, is carefully analyzed in this study, particularly in Chapters 6 and 7.

6. See Lane, *Architecture and Politics*, p. 103, and Borngräber, "Foreign Architects in the USSR," 1979, for the relationship of German architects to the Soviet Union.

7. The migrations of the German architects are derived from many sources: from biographies in the standard encyclopaedias; from Lane, *Architecture and Politics*, 1968; from Wingler, *Bauhaus*, 1969; and from Moholy-Nagy, "The Diaspora," 1965.

8. According to Jeremy Gould, *Modern Houses in Britain, 1919–1939*, 1977, p. 15, these émigré architects included Eugen C. Kaufmann, A. E. Proskauer, Ernö Goldfinger, Rudolf Frankl, Ernst L. Freud, and R. A. Ruheman. According to Dennis Sharp, *Sources of Modern Architecture*, 1967, p. 21, Korn attended a CIAM conference in London in 1934 with Walter Gropius, and then returned to England again in 1937, this time to settle.

9. Moholy-Nagy, "The Diaspora," 1965, argued that this falling off in creativity related not only in Germany but to the German architects in exile, cut off from the cultural roots which nourish them.

10. The development of modern architecture in Palestine (Eretz Israel) in the 1930s is a phenomenon not yet adequately studied and documented. For a broad overview, with bibliographical references, see Herbert, "Israel," in Sanderson, *International Handbook of Contemporary Development in Architecture*, 1981, pp. 351–359.

11. "Holzhausbau," *Wasmuth's Monatshefte*, 1930, pp. 564–565.

12. "Konrad Wachsmann, Berlin: Holzhausbau," *Moderne*

Bauformen, 1931, pp. 401–403, 418–419.

13. Wachsmann, "Kleiner und grosse Bauten in neue Holzbautechnik," *Bauwelt,* 1931, pp. 1559–1574.

14. He returned briefly to his home, to try and persuade his family to leave. Unfortunately, he could not convince them of the reality of the danger confronting them. They perished in the Holocaust.

15. This account is based in part on Wachsmann's unpublished autobiography and in part on various curricular vitae found in Wachsmann's papers. Wachsmann was not precise in his use of dates, and there are minor inconsistencies between the various documents.

16. These competition entries were illustrated and described in *Bauwelt,* 9, 27 February 1931, pp. 14, 43. The reports of the assessors, in manuscript form, may be found in the Gropius archives.

17. The use of this material by Gropius was cited in Rasch, *Wie Bauen?* 1927, p. 143.

18. The official Nazi party newspaper, edited by Rosenberg. Here cited by Lane, *Architecture and Politics,* 1968, p. 165.

19. Lane, *Architecture and Politics,* 1968, pp. 162, 165.

20. According to Lane, *Architecture and Politics,* 1968, p. 181, Gropius wrote to the *Reichskulturkammer* in both February and March 1934. Martin Wagner followed this with a strong letter, much less conciliatory in tone, in June.

21. Howard Robertson, "The First Nazi Exhibition, Berlin," *The Architect and Building News,* 18 May 1934, p. 194.

22. According to a letter from Walter Gropius to Barbara Miller (Lane), of 14 April 1955, in the Gropius archives.

23. "The Gropius Exhibition," *The Architect and Building News,* 18 May 1934, p. 181.

24. This view was expressed in a letter from Maxwell Fry to the author, 17 February 1982.

25. Gropius, "The Formal and Technical Problems of Modern Architecture and Planning," *Journal of the Royal Institute of British Architects,* 19 May 1934, pp. 679–694; also excerpts published in *The Architect and Building News,* 18 May 1934, pp. 203–206, from which I quote here.

26. Maxwell Fry, "Walter Gropius," *Architectural Review,* March 1955, p. 155.

27. Gropius, "The Formal and Technical Problems," 1934, p. 204.

28. Pritchard was a leading member of the Design and Industries Association, had edited its journal, and was active in the Isokon furniture venture, with which Gropius had connections while in England.

29. According to the account given by Ise Gropius in "History of the Gropius House in Lincoln, Massachusetts," 1977, p. 1.

30. James Marston Fitch, *Walter Gropius,* 1960, p. 23.

31. Wachsmann, "Zum 75. Geburtstag von Walter Gropius," *Baukunst und Werkform,* 1958, p. 284. This, and the autobiography, are our main sources for this meeting.

32. Gropius, "Theaterbau," *Convegno di Lettere, Octobre 1934,* 1935, pp. 154–162. The convention was held from 8 to 14 October 1934, and Gropius' address dealt with the development of the modern theater, with particular emphasis on his own "total theater" proposal.

33. It is not clear whether the German government expected Gropius to return to Berlin from Rome. Both Fry's invitation to Gropius to work with him, and several letters of Gropius (in the Gropius archives) suggest that the undertaking of a professional commission was the reason officially given for Gropius to leave Germany, in which case London was his announced destination. On the other hand, both Fitch and J. M. Richards, in *Who's Who in Architecture,* 1977, p. 129, suggest that Rome only was approved, and his journey on to London was unexpected by the Germans. Richards perhaps over-romanticizes the situation: "Gropius . . . got a permit to attend a theatre conference in Rome. He threw away his return ticket to Dessau, and traveled on to London. . . ." Gropius of course had long since left Dessau and had been living in Berlin for several years.

34. Maxwell Fry tells the following story of Gropius during those difficult days: "At one time when things were a bit pinched I grumbled, and he said to me, 'Do you love architecture, Max?' I said 'Yes.' 'Then you must pay for it.' " Gilbert Herbert and Sydney Abramowitch, "Interview with Maxwell Fry," *South African Architectural Record,* May 1946, p. 131.

35. According to Richards, *Who's Who in Architecture,* 1977, p. 129.

36. Early letters from Gropius stress that he has gone to England to undertake professional work and has not left Germany. This might indeed have been his intention, although it is unlikely; it is more probable that he is being cautious, not only for his own sake, but for his family, and Ise Gropius' family, who remained behind in Germany. Giedion, *Walter Gropius,* 1954, p. 10, puts his frame of mind thus: "A few years' break in England and then it might be possible to return."

37. The first approach was made through Alfred Barr, of the Museum of Modern Art, at Hudnut's request, in the summer of 1936. In December, Gropius accepted, and the formal announcement was made on 15 January 1937. ("Modern Architecture Symposium: The Decade 1929–1939," *Journal of the Society of Architectural Historians,* 24:1, 1965, p. 93.) Hudnut had previously contacted Oud in Rotterdam and Mies van der Rohe in Berlin, as possible candidates (Ise Gropius, "History of the Gropius House," 1977, p. 1).

38. Ironically, it had been Jaeckh's contacts with Hitler and Rosenberg, in March 1933, that had prompted Gropius and Wagner to resign from the Executive Council of the Werkbund, in protest. Lane, *Architecture and Politics,* 1968, p. 174.

39. Ise Gropius, "History of the Gropius House," 1977, p. 1.

40. For the more personal elements of Wachsmann's life,

we must await the publication of his autobiography. The facts of his troubled sojourn in France are based on my reading of the draft manuscript of that story.

41. Letter Walter Gropius/ French Ambassador to Italy, 18 March 1940.

Chapter Seven

1. Kenneth Kingsley Stowell, "Housing the Other Half," *Architectural Forum* 56:1932, pp. 217–220.

2. Nathaniel S. Keith, *Politics and the Housing Crisis since 1930*, 1973. According to Keith (p. 22) by 1933 there were between 12 and 17 million persons unemployed in the United States.

3. An indication of the widespread use of pre-made components may be derived from a number of articles in *Bauwelt* in 1931, by Jobst Siedler, including a long series dealing with building innovations at the *Bauwelt* sponsored Berlin building exhibition of 1931, entitled "Neues Bauen," in issues numbers 26, 28–36, 38, 39, 42, 45–48, 50, 51.

4. For examples of planned housing estates in Philadelphia, Cleveland, and New York, see Richard Pommer, "The Architecture of Urban Housing in the United States during the Early 1930s," *Journal of the Society of Architectural Historians*, 37:1978, pp. 235–264.

5. For example, Alfred Kastner and Oskar Stonorov, in Philadelphia, who "had in mind German models such as the Dammerstock Siedlung of Gropius and Haesler in Karlsruhe, Gropius' for Spandau-Haselhorst, or Schumacher's work in Hamburg . . . " (Pommer, "The Architecture of Urban Housing," 1978, pp. 239–240); or Wallace Teare and William Conrad of Cleveland, who "began research on concrete slab systems for housing under the inspiration of Ernst May's work in Frankfurt." (Ibid., pp. 246–247.)

6. Catherine Bauer, *Modern Housing*, 1934, p. 240.

7. The "other half" is a convenient but probably misleading abstraction, used to dramatize a situation which, by the mid-thirties, was such that "some 79% of American families could not afford a 'low-cost' house priced with lot at $4,000." (*Family Expenditures in the United States*, National Resources Planning Board, Washington, 1941, cited by Kelly, *The Prefabrication of Houses*, 1951, p. 29.)

8. Catherine Bauer Wurster, "The Social Front of Modern Architecture in the 1930s," 1965, p. 52.

9. Kelly, *The Prefabrication of Houses*, 1951, pp. 28–29.

10. Ibid., p. 49. For details of the development of prefabrication in this period, there is still no more comprehensive account, well documented and thoughtfully analyzed, than this of Kelly's.

11. Emmet, *Catalogues and Counters*, 1950, p. 521, cites Arthur F. Burns, *Production Trends in the United States Since 1870*, New York, National Bureau of Economic Research, 1934, pp. 302–303.

12. In addition to Kelly, information on prefabrication firms comes from published lists of firms, such as "Prefabrication Gets Its Chance," *Architectural*

Forum, February 1942, especially "Snapshot of an Infant Industry," pp. 84–88; "Prefabrication Up-to-Date," *American Builder*, January 1943, especially "Directory of 'Prefabricators Up-to-Date,' " pp. 45, 77–79; and "A List of Prefabricators," *Architectural Record*, June 1943, pp. 75–79.

13. Emmet, *Catalogues and Counters*, 1950, pp. 226–228, 520–530.

14. See J. L. Peterson, "History and Development of Precast Concrete in the United States," *Journal of the American Concrete Institute*, February 1954, pp. 483–486, for an account of Conzelman's contribution.

15. *Engineering News*, 24 January 1916, cited in Peterson, "History and Development of Precast Concrete," 1954, p. 486.

16. See the survey of these systems in Bruce and Sandbank, *A History of Prefabrication*, 1945, ch. 3.

17. Prentice Winchell, "The Dwelling of Tomorrow: An Economic Study in Residential Construction Showing Why a Growing Use of Iron and Steel is Inevitable," *Iron Age*, 1926, pp. 613–615, 686–687, 766–768, 840–841, 930–931, 992–993.

18. Winchell, "The Dwelling of Tomorrow," 1926, p. 766.

19. Winchell, "The Dwelling of Tomorrow," 1926, p. 841.

20. "Steel Houses in Germany," *Iron Age*, April 1928, p. 934.

21. Gropius Copper Houses were reported in "Portable Houses of Copper," *Architectural Forum*, February 1933, pp. 145–147.

22. See "Portable Steel and Copper House Developed in Germany," *Iron Age*, February 1931, p. 548; "Germany Now Has Copper Houses Made in Sections Easily Set Up", *Copper and Brass Bulletin*, 1931, pp. 2–3.

23. The Naugle House for the Tuxedo Park Association, cited in Bruce and Sandbank, *A History of Prefabrication*, 1945, p. 42.

24. Spiegel, *Das Stahlhausbau*, v. 1, 1928, pp. 29ff.

25. See reports in *Architectural Forum*, 56, March 1932, pp. 285–288; *Fortune*, 6, July 1932, pp. 64–65.

26. "Beech Aircraft to Use Reconverted Plants for New Type Home Manufacture in Cooperation with Dymaxion Dwelling Machines, Inc.," *Aviation News*, 2, 20 November 1944, p. 34. Fuller was Chairman and Chief Engineer of the Corporation from 1944 to 1946.

27. "Fuller's House Has Better Than Even Chance of Upsetting the Building Industry," *Fortune*, 33, April 1946, pp. 166–172.

28. "What Became of the Fuller House?" *Fortune*, 37, May 1948, p. 168.

29. Llewellyn and Speller, "Steel in Residence Construction," *Architectural Record*, June 1933, p. 441. This was a steel-framed house, covered in metal shingles. Rowley also designed a steel house for the Wheeling Corrugating Company (both houses illustrated in "Technical News and Research," *Architectural Record*, January 1934, pp. 21, 25).

30. Noted in *Architectural Forum*, October 1932, sup. p. 20.

31. "Steel Houses," *Architectural Forum*, April 1933, p. 330. The article notes that "U.S. Steel Corporation has studied prefabricated houses, apparently has built none."

32. The origin of General Houses is given in "Steel Houses," 1933, pp. 330–331; also in Kelly, *The Prefabrication of Houses*, 1951, p. 39.

33. For American Houses see "Steel Houses," 1933, pp. 328–329; and Kelly, *The Prefabrication of Houses*, 1951, p. 41. American Houses were the producers of the widely publicized Motohome, with a mechanical core designed by the Pierce Foundation and General Electric. In 1938 American Houses abandoned the steel frame, and produced wood-framed structures in large numbers.

34. See Kelly, *The Prefabrication of Houses*, pp. 49–50.

35. "Steel Houses," 1933, p. 327.

36. "Steel Houses," 1933, p. 327.

37. See press cuttings from the *New York Times*, 25 May 1932, 23 June 1932, 3 December 1933; and *Time*, 4 July 1932, 27 March 1933, reproduced in Bruce and Sandbank, *A History of Prefabrication*, 1945, p. 7.

38. "The Modern Houses of the Century of Progress Exposition", *Architectural Forum*, July 1933, pp. 51ff.

39. As an example of the detailed coverage given to prefabricated housing in the professional press, see "Technical News and Research: New Housing Designs and Construction Systems," *Architectural Record*, January 1934, pp. 12–13.

40. For a review of this research activity, see Bruce and Sandbank, *A History of Prefabrication*, 1945, pp. 10–14. "Non-commercial Research and Experimentation"; Kelly, *The Prefabrication of Houses*, 1951: pp. 21–25, "Research by Bemis," pp. 31–32, "Non-commercial Research and Development," p. 33, "US Forest Products Laboratory"; Burchard, "Research Findings of Bemis Industries, Inc.," *Architectural Record*, January 1934, pp. 3–8. For Pierce Foundation, see *Architectural Forum*, 72, May 1940, pp. 365–369.

41. See Bruce and Sandbank, *A History of Prefabrication*, p. 14; Kelly, *The Prefabrication of Houses*, 1951, pp. 36–37.

42. Keith, *Politics and the Housing Crisis since 1930*, 1973, p. 35.

43. Keith, *Politics*, 1973, p. 39.

44. "Dr. Gropius Looks into Future, Sees Better World Tomorrow," *Boston Herald*, 7 February 1939.

45. Keith, *Politics*, 1973, p. 39.

46. Letter Martin Wagner/Walter Gropius, 8 September 1940. Wagner (1885–1957) joined the Harvard Design School faculty in 1938, and in 1950 retired as Associate Professor of Regional Planning, Emeritus.

47. Wagner, *Das Wachsende Haus*, 1932.

48. Martin Wagner, *Prefabricated Housing*, unpublished manuscript (typescript amended by hand for publication) written in 1939–40, foreword dated 6 July 1941. Drawings (missing from manuscript) ascribed to Bernhard Wagner and Heinrich Shapiro, translation assisted by Oscar Sutermeister.

49. Wagner, *Prefabricated Housing*, 1941, pp. 71–76, analyzes the work of Gropius and Bartning at the Berlin Growing House exhibition and, despite some pertinent technical criticism, concluded that both were working in the right direction in using high-technology metal systems.

50. See *Architectural Forum*, 74, February 1941, pp. 87–90, for a full description of the igloo house.

51. Joint publications by Walter Gropius and Martin Wagner: *Cities Renaissance*, 1942; *Housing as a Townbuilding Problem*, 1942; *New Boston Center*, 1942; "The New City Pattern for the People and by the People," *Conference on Urbanism: The Problems of the Cities and Towns*, 1942.

52. Walter Gropius and Martin Wagner, "How to Bring Forth an Ideal Solution of the Defense Housing Problem?" in *U.S. 77th Cong., 1st sess., House Select Committee Investigating National Defense Migration*, 1941, H. Doc. 17, pp. 6949–6956.

53. Ibid., p. 6956.

54. "Exhibit 2—Text of Lanham Act, as Amended," *U.S. 77th Cong., 1st sess., House Select Committee*, p. 6956.

55. "Building for Defense," *Architectural Forum*, 75, August 1941, p. 107.

Chapter Eight

1. According to Wachsmann's draft autobiography.

2. Originals and prints of ten of these drawings were among Wachsmann's drawings at the University of Southern California, and had been identified by Wachsmann, to Elizabeth Wang, who had been in the process of cataloging his drawings, when he died. It was during that process, early in 1981, that I had the opportunity to examine the drawings. In 1982 I received a draft of the index of all Wachsmann's works, which was produced for microfilming by the Huntington Library, in cooperation with the Smithsonian Institute. The index, compiled by Robertson Ward, Stephen Cruz, and Elizabeth Wang, under the guidance of Judith Wachsmann, is entitled: *The Complete Project Works of Konrad Wachsmann, May 16, 1901–Nov. 25, 1980*. In this book the numbers of drawings in various sets are those which derive from my initial survey, and not from the index. In his autobiography, Wachsmann says he arrived with 13 drawings.

3. Draft autobiography, Wachsmann.

4. Ibid. The Florida project was published in *Architectural Forum*, 77, August 1942, pp. 83–85.

5. Wachsmann talks of the "irrevocable break" with Breuer in his autobiography. From the content of an exchange of letters between Gropius and

Breuer in 1941, it is apparent that a bitter dispute in the architectural school had spilled over and corroded their professional relationship.

6. Draft autobiography, Wachsmann.

7. Walter Gropius and Martin Wagner, "How to Bring Forth an Ideal Solution of the Defense Housing Problem?" in *U.S., 77 Cong., 1st sess., House Select Committee Investigating National Defense Migration*, 1941, H. Doc. 17, pp. 6949–6956.

8. The provenance of this set of drawings (the Modified French scheme) poses a problem. The fact that all drawings are signed by Wachsmann alone, and that Gropius' name does not appear, would seem to indicate that they were prepared before 7 December 1941, when the decision to collaborate was reached. If this is correct, it is strange that this was not the set Wachsmann showed to Gropius. After all, it was much more developed, with an improved joint and coupling system. On the other hand, if these drawings were prepared after Pearl Harbor, then they were completed in three weeks (before the end of 1941), a considerable feat, considering the extent of redesigning involved. In the latter case one can only assume that Gropius' name is missing because no agreement had yet been reached as to the style of signing the drawings. From this time onward, the names of both partners appear on all drawings, generally with the name of that partner first who was primarily responsible for the particular drawing in question. In the nature of things, therefore, most drawings are labeled "Konrad Wachsmann and Walter Gropius."

9. Letter Wachsmann/Kurt Friedberg, 2 February 1942. According to the portfolio of drawings, there were only 22 drawings in all, plus a cover.

10. Ibid.

11. According to the draft autobiography. Kurt Friedberg may have been the first, as in his letter to Friedberg (2 February 1942), Wachsmann indicated his intention of sending a full set to London, for Friedberg's critical comment.

12. The cover of the portfolio read: "The Packaged House System by Konrad Wachsmann and Walter Gropius 1942, Cambridge, 22.2.42."

13. The term "packaged house" had been used earlier by F. Vaux Wilson (a pioneer prefabricator of the Homasote Co. of New York) in his article entitled: "New Approach to 'Packaged' House Recognizes Architects," *Architectural Record*, 82, August 1931, pp. 86–87.

14. Patent Serial No. 445216, "Prefabricated Building," application filed by Konrad Wachsmann and Walter Gropius, Lincoln, Mass., 30 May 1942.

15. These were wall-size panels, but Gropius proposed reducing them to modular panels 1 m wide.

16. Not only in the professional journals, especially *Bauwelt* and *Deutsche Bauzeitung*, but in the book: Martin Wagner, *Das Wachsende Haus*, 1932. Specific examples using a load-bearing panel system include Gascard and Canthal, and Kohler and Schweitzer. In both cases the framed panels are bolted together.

17. "Modulok," *Architectural Forum*, 79, September 1943, pp. 65–68. As no date is given for the development of the system, priorities cannot be established.

18. D. Dex Harrison et al., *A Survey of Prefabrication*, 1945.

19. Ibid.

20. Draft autobiography.

21. One of the drawings showing the *neue Wandverbundengung* is dated 8.1.32, and signed Goetz (Gropius' assistant). A scheme analogous to the original Hirsch detail (of fixing the panels to a light metal post) was that of Otto Bartning, developed in conjunction with an aircraft manufacturer. Here the steel studs are triangular, and there is a rubber filler between wall panel and stud. See Wagner, *Das Wachsende Haus*, 1941, pp. 71–72.

22. Letter Gropius/John Burchard, 19 February 1942.

23. Ibid.

24. He had built his own house in Lincoln, and other residences, in timber and, together with Marcel Breuer, had constructed a large defense housing scheme in New Kensington, all of wooden construction—published in *Architectural Forum*, 75, October 1941, pp. 218–220.

25. In a folio entitled "The Packaged House. Walter Gropius and Konrad Wachsmann. Copyright 1942 by Walter Gropius and Konrad Wachsmann," containing 5 out of 6 planned sheets.

26. Konrad Wachsmann, "Machine Energy: The Technique for our Time," in T. H. Creighton, *Building for Modern Man*, 1949, pp. 46–48.

27. Walter Gropius, "Prefabrication: A Freedom from Limitations," in Creighton, *Building for Modern Man*, pp. 41–45.

28. Cited in Oscar Handlin, *Truth in History*, 1979, p. 321.

29. In many ways Gropius' friends and contacts proved helpful to the development and furtherance of the packaged house. John Burchard, of the Bemis Foundation, wrote to the Office of the Chief of Engineers, U.S. Army (February 1942) highly recommending Gropius' prefabricated system; Gropius' position on the Advisory Committee to the Government on Prefabrication gave him access to Washington, as did his friendship with Reginald Isaacs, then working for the Defense Housing Department; his friends Siegfried Giedion, José Luis Sert, and Aldous Huxley not only lent their prestige to the project but supported it by investing in its development (Wachsmann, autobiography).

30. Not only was this a psychological burden but a real physical handicap. Not only were they not allowed to fly, but all travel was restricted, and even a rail journey from New York to Boston required a permit.

31. The growing tension between Ise Gropius and Konrad Wachsmann is noted in Wachsmann's autobiography, and was also stressed by Mrs. Gropius, in an interview with the author (1 April 1981).

32. Nathaniel S. Keith, *Politics and the Housing Crisis since 1930*, 1973, p. 40.

33. "Prefabrication Gets Its Chance," *Architectural Forum*, 76, February 1942, pp. 81–88.

Chapter Nine

1. As Wachsmann wrote to Kurt Friedberg (24 July 1942): "Das wesentliche, wenigstens fuer mich ist dass ich eine neue Konstruktionsmethode ausgeknobelt habe (bitte lachen Sie nicht) die ich fuer viel besser halte als die Alte."

2. Wachsmann's English was at this time very poor, although he probably knew more than the one sentence he had claimed to master: "A thunderstorm refreshes the air" (autobiography).

3. Certificate of incorporation, General Panel Corporation, 12 September 1942.

4. Letter Wachsmann/Friedberg, 7 September 1942. This is an elated letter, beginning with the English phrase "Good News today"!

5. Letter Gropius/Burchard, 4 February 1943. The demonstration had been planned for 19 February, and then deferred to 16 February, as this letter states. It eventually took place on 23 February 1943.

6. Sectionalized Construction for Temporary Dwelling Units, 9 January 1943 (7 sheets); Sectionalized Construction for Temporary Dwelling Units, 5 June 1943 (10 sheets); Sectionalized Construction for Military Barracks, n.d. (7 sheets).

7. Certificate of increase of Number of Directors of General Panel Corporation, filed 15 March 1943; Certificate of Number of Shares and Change of Statement Respecting Capital of General Panel Corporation, filed 15 March 1943. The documents are signed by Charles Allen Sr., Leon R. Spear, Alexander Louria, M. L. Heide, Jack Marqusee, David G. Baird, Konrad Wachsmann, and Walter Gropius. Marqusee was designated secretary.

8. Willo von Moltke, "Prefabricated Panels for Packaged Buildings," *Architectural Record*, 93, April 1943, pp. 50–53.

9. Herman Herrey, "At last we have a Prefabrication System which enables Architects to Design any type of Building with 3-dimensional Modules," *Pencil Points*, 24, April 1943, pp. 36–47.

10. "Prefabricated Houses," *Architectural Forum*, 77, June 1943, pp. 89–96, with special reference to General Panel on pp. 94–95.

11. *Christian Science Monitor*, 23 February 1943, p. 2; *Colliers*, 12 June 1943 (advt. "Miracle House That Grows with Your Family"); *Business Week*, 11 December 1943.

12. The possible exception here is the erection of the demonstration house in New York. The references to this are oblique. Sert wrote to Gropius, on 19 August 1943, saying "Wachsmann phoned me a few days ago saying that you are bringing your model to New York and wanted to set it up in some available lot in the city right away." Presumably the model referred to is the full-size demonstration house. *The Business Week* article of 11 December 1943 describes the General Panel house erected in New York (behind a stout fence) to test the weathertightness.

13. Interview with Rudy Wolf, 17 May 1981.

14. "Variety of Houses from Identical Prefabricated Units of General Panel Corp.: Designed by Harvard Students," *Pencil Points*, 24, December 1943, pp. 76–77.

15. Set of blueprints labeled "The Flexible House by Walter Gropius. GP Corp. Aug.–Sept. 44."

16. "Expansible Prefab House for Postwar. Walter Gropius, Architect," *Architectural Record*, 96, December 1944, p. 69.

17. See, for instance, the documents for a two-bedroom cottage, prepared by Wachsmann, with 11 sheets of drawings (10 January 1944), schedule and specification of standard panels.

18. Application 29 December 1943, serial no. 516050, patent 2,426,802.

19. According to General Panel Corporation Prospectus, 5 February 1946, Gropius was committed until 17 September 1947, and Wachsmann until 17 September 1949.

20. Drawings and photographs in Wachsmann files. Gropius had a partition plan for the General Panel Corporation offices at 103 Park Ave., in New York, presumably of the new system, dated 26 April 1944. The partition system and the house were exhibited in an elegantly designed set of displays in Philadelphia, in 1944.

21. According to General Panel Corporation Prospectus, 5 February 1946.

22. The Mobilar system was featured in 1944 in a one-man show at the Museum of Modern Art in New York *Mobilar Structure*, with an introduction by Le Corbusier. For notes on the system, see Robertson Ward, "Konrad Wachsmann: Towards Industrialization of Buildings," *AIA Journal*, March 1972, pp. 33–43, and Konrad Wachsmann, *The Turning Point of Building*, 1961.

23. Prior to the war, in July and August 1939, there was a flurry of correspondence between Wachsmann, at Vence, and Kurt Friedberg, in London, in an attempt to get the system of hangar construction patented.

24. The drawings are labeled: " 'Mobilar' hangar system, Atlas Aircraft Products Inc., Hangar Division, New York. Invented by Konrad Wachsmann, March 1945."

25. Patent 2,355,192, 8 August 1944.

26. Two similar but not identical sets of drawings exist, the one labeled "Atlas Aircraft Products Corp., N.Y.–L.A. (dated 7 and 8 August 1944, approved: Wachsmann)"; the other labeled "General Panel Corp., N.Y. (dated from August 1944 to January 1945)."

27. Patent filed 10 August 1945, serial no. 609996, granted 27 May 1947, no. 2,421,305.

28. The original design, and the demonstration houses, used external wooden siding as the finish to the panel. Influenced by the experimental work in plywood of the John B. Pierce Foundation, they switched to stressed skin plywood panels.

29. Prospectus, 5 February 1946.

30. Keith, *Politics and Housing*, 1973, p. 43.

31. The *Architectural Forum*, April 1943, listed 71 wartime prefabricators; the number had grown, by April 1945, to 95, according to NHA sources. In the *Architectural Record*, June 1943, the following are included in firms producing more than 1000 units per month: Aladdin Co.; American Houses; Bates Prefabricated Structures; Houston Ready-Cut Houses; National Homes. The total figures of wartime production are derived from Kelly, *The Prefabrication of Houses*, 1951, p. 60.

32. Production did not really depend on acquiring a fully equipped factory, although this was Wachsmann's dream. Until factories were built, it was reported, "subcontracting firms will be employed, and houses will be turned out. . . ." *Architectural Forum*, 84, February 1946, pp. 7–8.

33. Ibid. There had been a previous limited licensing arrangement with the Burton Rodgers Corporation (vide Prospectus) which apparently did not lead to practical results.

34. Harrison, *A Survey of Prefabrication*, 1945.

35. Draft autobiography.

36. According to information supplied to me by Reginald Isaacs, Gropius had complained of these costs, and especially of high costs of entertainment, travel, etc., which the corporation had incurred.

37. According to the balance sheet of the corporation of 31 December 1945, included in the Prospectus of 5 February 1946.

38. Undated letters Weissberger/Gropius (the Mexico letter is dated 8.8.44).

39. In mid-1944, according to a certificate filed in the New York Department of State, on 5 May 44, Marqusee was shown as president of General Panel; in a certificate of 23 January 1946, Wachsmann appears as president. This change may indicate Allen and Co.'s withdrawal from General Panel. In the list of directors in the Prospectus of 5 February 1946, Wachsmann is shown as president; Charles Wohlstetter, chairman of the board, treasurer and director; Leon R. Spear, vice-president and director; Alexander L. Louria and Albert Wohlstetter, directors. Roy Plaut is secretary, and Nathan K. Mendelsohn, assistant treasurer.

40. Curtis Fremond was Wachsmann's old friend and adviser, who had come from London to join him; Rudy Wolf was described as "chief draftsman" in a trip report of the Bemis Foundation, 14 October 1946.

41. In the Wachsmann tradition, all fully (almost compulsively) detailed, showing every element with alternatives for different specifications; dated 1945.

42. In a report by C.R.F. (Fremond) dated 16 July 1945: "Miss Weed was kept busy . . . completing the books for Sweden and Canada."

43. Certificate of Authorization of Shares . . . , etc., General Panel Corporation, filed 23 January 1946.

44. Prospectus, 5 February 1946.

45. Keith, *Politics and Housing*, 1973, p. 59.

46. Ibid., p. 60. 1,023,000 units were actually produced in 1946, for middle-class users rather than the working class.

47. Most contemporary references locate the experimental houses at Whitestone, Long Island. Mr. Wolf, in his interview of 17 May 1981, says the houses were at Astoria, Long Island. Both sites are close to, but on opposite sides of, La Guardia Airport.

48. Houses by Neutra, Gropius, and Elsa Gidoni are published in *Prefabricated Homes*, November–December 1946, p. 27; houses by Wachsmann, Gropius, and Neutra in *Architectural Forum*, 84, February 1946, pp. 7–8; there is a plan for a two-bedroom house by Richard Neutra, under the General Panel Corporation imprint, undated, in the Wachsmann files.

49. *Architectural Forum*, 84, February 1946, pp. 7–8.

50. Letter Walter Gropius/Konrad Wachsmann, 20 July 1945.

51. He kept a set of prints of this scheme (dated February–April 1942) together with his records of the Packaged House.

52. Illustrated in "Prefabrication Gets Its Chance," *Architectural Forum*, 76, February 1942, p. 31; briefly described in Kelly, *The Prefabrication of Houses*, 1951, p. 32; *Architectural Forum*, 74, November 1941, pp. 321–326.

53. *Architectural Record*, 93, June 1943, p. 76. An example of the use of Cemesto, for the Arlington Farms Women's Dormitories, near Washington, D.C., is illustrated on pp. 68–69.

54. See note on Dahlberg's views presented to the Senate's Postwar Economic Policy and Planning Commission, reported in "Postwar Promise," *Architectural Forum*, 79, September 1943, pp. 41–42.

55. Letter Carl Dahlberg/Burnham Kelly, 13 November 1946; trip report on General Panel, Bemis Foundation, 20 February 1947.

56. *Celotex News*, October 1943.

57. These discussions generated a rumor that Celotex was planning to take over General Panel, which was reported to Burnham Kelly and noted on 21 May 1946.

58. Many reports on the Celotex connection noted with some surprise that Celotex products were not used in the General Panel house. See, for instance, *Architectural Forum*, 86, February 1947, p. 115: "The Celotex Corp.—despite the fact that there is not so far an ounce of Celotex products in the house—will distribute, retail and erect it." Earlier, the *Architectural Forum*, 86, January 1947, suggested that this situation might change in the future: "No Celotex products are now used in the General Panel houses, but Wachsmann is at work on a metal house which may incorporate some Celotex products."

59. For the setting up of General Panel Corporation of California, information has been culled from many sources, including: Standard and Poor's *Standard Corporation Records*,

February–March 1948, p. 879; Dun and Bradstreet Report on General Panel Corporation of California, 16 December 1948; Bemis Foundation Trip Reports, General Panel Corporation, 14 October 1946 and 20 February 1947.

60. In the well-documented account of the California firm in *Arts and Architecture*, 64, November 1947, pp. 28ff., Wachsmann does not appear among the officers of the company; on the other hand, in *Architectural Forum*, 86, February 1947, he is referred to a vice-president. In the authoritative Dun and Bradstreet report of December 1948, Wachsmann is listed as vice-president of the California corporation; official California Corporate records are no longer available.

61. In a letter to Secretary Roy Plaut, 22 August 1946, Wachsmann informed him that he had decided not to participate financially in the California company; he was angry that he had not been consulted adequately, and foresaw a conflict of interest with his position in New York; in a letter to Plaut, 17 January 1947, he again complained about lack of adequate discussion with him and of conflicts between California and New York.

62. Details of the acquisition of the Burbank factory derive from the Dun and Bradstreet report of 16 December 1948; *Architectural Forum*, 86, January 1947; and draft autobiography, Wachsmann. Sizes given for the factory vary from 225,000 to 250,000 sq ft. The cited figure is that given by Wachsmann himself.

63. *Architectural Forum*, 86, January 1947.

64. Letter Charles Wohlstetter/Konrad Wachsmann, 12 June 1946.

65. A booklet produced by the Market and Cost Research Department of General Panel, 12 August 1946, indicates the cost (before dealers profit) of the "Minimum House" as $4,165, of which the shell cost (including $442.50 overheads and profit) was $1,665.

66. Letter C. B. Dahlberg/National Housing Agency, 27 September 1946.

67. Letter Office of Housing Expediter/Carl Dahlberg, 20 December 1946.

68. Keith, *Housing and Politics*, 1973, p. 67.

69. *Architectural Forum*, 86, January 1947. Gropius responded to these negative developments with a letter to the editor of the *New York Times*, 2 March 1947, published under the heading "Progress in Housing," in which he drew attention to the serious effect of Wyatt's resignation and other steps.

70. Letter Dahlberg/Burnham Kelly, 13 November 1946; trip reports, Bemis Foundation, 14 October 1946, 20 February 1947.

71. Letter Wachsmann/Roy Plaut, 17 January 1947. Mechanical engineering drawings of machinery under the General Panel imprint, were prepared by L. Hall, during April and May 1947.

72. Trip report, Bemis Foundation, 20 February 1947.

73. *Architectural Forum*, 86, January 1947.

74. Trip report, Bemis Foundation, 20 February 1947.

75. Trip report, Bemis Foundation, 10 April 1947; letters General Panel of California/American Manufacturing Co., 30 April 1947, and Radio Corporation of America/Wachsmann, n.d.

76. In his autobiography Wachsmann says he realized that the "whole principle [was] wrong." Just before he died, at the end of 1980, Wachsmann told Rudy Wolf that the concept of a four-way connector as a universal joint was not really necessary, and that things could work out equally well with a two-way joint (interview with Wolf, 17 May 1981). Wachsmann was interested in alternatives. We have already observed that he was investigating a metal house at the beginning of 1947.

77. Letter from Shapiro/Wachsmann, 15 June 1947, referring to presentation drawings Shapiro was preparing of the factory layout.

78. Discussed in memo to General Panel management from Albert Wohlstetter, 20 June 1947.

79. Dun and Bradstreet report, 16 December 1948.

80. Pamphlet VIII: "Technical Performance," General Panel Corporation.

81. It was the original intention to set up a separate organization in Oregon to supply, from their own or leased timber stands, milled parts ready for use. Trip report, Bemis Foundation, 20 February 1947.

82. There was a controversy over the use of radio frequency heating rather than by steam platens, which Albert Wohlstetter regarded as more reliable. Letter A. Wohlstetter/author, 23 June 1981. An alternative exterior to plywood, namely the use of redwood siding (as in the early experimental houses) was an option which remained open. Pamphlet VIII: "Technical Performance," General Panel Corporation.

83. Problems with the unions placed restrictions on mechanized methods in traditional crafts. For instance, spray painting in the workshop was at first restricted by the union to clear sealers only (later, a prime and second coat applied in the shop were permitted). As a further example, welded steel plumbing assemblies, which would have been more economical and efficient, were prohibited by the Master Plumbers' Union. Trip report, Bemis Foundation, 20 February 1947.

84. Pamphlet: "The General Panel System. Invented by Konrad Wachsmann and Walter Gropius."

85. Trip report, Bemis Foundation, 20 February 1947.

86. According to Wohlstetter, the wasted 8" per sheet cost General Panel about 20 percent more on material (letter 23 June 1981). Southwell was anxious to have much larger wall units, as the houses were standardized in any event. (One can imagine how Wachsmann and Gropius would react to this adulteration of the modular purity of their scheme.) Of course, special 40" sheets of plywood might have become available economically, if the production had ever reached the volume. (10,000 per an-

num per shift) planned. As far as the redundant framing was concerned (and in order more easily to maintain the fine tolerances required), the alternative of steel framing to the panels was being explored, according to the Trip report of 20 February 1947.

87. According to Josef van der Kar, in letter to author, 6 July 1981.

88. These filler strips created problems internally, where flush joints were considered desirable, and various solutions (taped joints, textured paint) were attempted. Evidence on the satisfactory quality of these solutions is contradictory (Wolf says the joint wore well, Van der Kar says the filler strips were loose on dry days).

89. The system attracted many architects, who were anxious to employ it for specific jobs. Vide one William Ferguson (letter to General Panel, New York, 8 April 1947) whose brother, an architect, wished to put up a General Panel house at Ithaca; or Howard Moise (letter to Konrad Wachsmann, 31 March 1947) who wished to use the General Panel system for two or three houses for his clients. As a professor of architecture at the University of California, Moise thought the venture would be an educational experience for his students.

90. Brochure: "Your General Panel Home," n.d.

91. Letter Shapiro/Wachsmann, 15 June 1947.

92. Southwell anticipated that about 5 different models would be produced in 1947, working up to "a line of houses offered in 14 different models in the year to follow, varying from 600 to 1100 sq ft." Trip report, Bemis Foundation, 20 February 1947.

93. Trip report, Bemis Foundation, 20 February 1947.

94. General Panel estimated that a trained crew could erect a house in 35 man hours (trip report, 20 February 1947). Some difficulties were experienced on the site. Four-way connections could prove tricky (Wolf) and there were problems with expansion on wet days (van der Kar).

Chapter Ten

1. "The Industrialized House," *Architectural Forum*, 86, February 1947, pp. 115–120; "House in 'Industry,'" *Arts and Architecture*, 64, November 1947, pp. 28ff.

2. Applications for appointments were received from Norman J. S. De Wind, formerly of the Technical Office of the Housing Agency (20 August 1947); Marc Vosk, market analyst and economist, who had been assistant regional expediter and research director in the New York office of the Housing Expediter (7 July 1947); Henry Soskin, formerly chief of community action program of the Office of Housing Expediter (8 July 1947); Walter Dushinsky (who stated he was a former Bauhaus student), technical manager and chief architect of the Commercial Aircraft Co., Tel Aviv, 1931–38, and connected with the design of the AIROH aluminium house in England (28 June 1947); J. van der Kar, who had in 1927 been associated with Buckminster Fuller's "Shelter" group (31 March 1947).

3. For example: Douglas Haskell, "Assembly Lines Reach Out for Markets," *Architectural*

Record, 93, June 1943, pp. 62–69; Fowler Manning, "Selling the Prefabricated House?—Here Are Some of the Unsolved Problems," *American Builder*, 66, February 1944, pp. 51ff; "Prefabrication," *Architectural Forum*, 84, April 1946, pp. 137–142, 190.

4. Salaries, according to an undated letter from Curtis Fremond to Wachsmann, were not being paid, and he looked forward to the time when the New York firm "will be solvent again." Ruth Wilson, a former employee, writing to Wachsmann (22 July 1947): "I gather from comments in the NY office that things are very unsettled throughout the GP organization."

5. According to Rudy Wolf (interview 17 May 1981).

6. In 1952 Wachsmann directed his students in a project (Shelter Design section) leading to the development of a wall panel of plastic sheets with a corrugated paper core, reinforced at the edges with an aluminium profile (for illustration, see Wingler, *Bauhaus*, 1969, p. 608).

7. The appellation "genius" crops up frequently in describing Wachsmann in informed communications: vide Wolf, Wohlstetter.

8. Kelly, *The Prefabrication of Houses*, 1951, p. 159. Examples given are Gunnison Homes (U.S. Steel Corp.); Wingfoot Homes (Goodyear Tire and Rubber Co.); Lustron (Chicago Vitreous Enamel Product Co.).

9. Dun and Bradstreet report, 16 December 1948.

10. Letter Wohlstetter/author, 23 June 1981.

11. Draft autobiography, Wachsmann.

12. Letter Walter Gropius/Helena and Simon Syrkus, 20 April 1948.

13. Many reports cite this stock of parts. See, for instance, that of Tom Aiken to the Bemis Foundation, March 1948; Nathan Mendelsohn, in interview (16 May 1981) noted that production was uneven, with the production of problematic components deferred while large stocks of the more straightforward panels accumulated; in the balance sheet of December 1947, merchandise held amounted to $181,000.

14. Mendelsohn confirms this view, in the interview of 16 May 1981.

15. See Trip report, Bemis Foundation, 20 February 1947. In June 1947 Wachsmann received a letter from David M. Goodman, who offered the opinion, in the spirit of constructive criticism, that the General Panel houses were much too expensive and would be at a cost disadvantage in relation to conventional building. He went on: "From what I have learned, your General Panel system is a real contribution. It would be tragic if it failed because of much the same reasons that the Kaiser Community Homes program has failed." Letter Goodman/Wachsmann, 10 June 1947. Van der Kar retrospectively confirmed this, when he suggested that General Panel failed, because it could not compete with cheaper tract hours (letter to author, 6 July 1981).

16. Details of reorganization from Dun and Bradstreet report, 16 December 1948; letter

Albert Wohlstetter/author, 23 June 1981.

17. Wohlstetter had served for years as factory manager for the Atlas Aircraft Products Corporation, and as consultant to the War Production Board and the Cyclohm Motor Corporation (Dun and Bradstreet report, 16 December 1948).

18. This moratorium was later renewed from time to time, by arrangement with the management.

19. Dun and Bradstreet report, 16 December 1948.

20. Dun and Bradstreet report, 16 December 1948.

21. Trip report, Bemis Foundation, 20 February 1947.

22. Letter Near Orient Agency (Eng. A. Averbuch)/General Panel, California, 25 April 1947. Averbuch was known to Wachsmann. Mendelsohn (interview 16 May 1981) recalls optimistic discussions of very large exports to Israel.

23. Letter Rudolf Hillebrecht/ Walter Gropius, 23 October 1948.

24. Interview with Rudy Wolf, 17 May 1981. The Wolf house, still in excellent condition, is at 2861 Nichols Canyon Rd., Los Angeles; interview with Nathan Mendelsohn, 16 May 1981; for van der Kar's house (house for Mr. and Mrs. Ralph C. Merten of Pacoima), see "This Factory Built House Came True," *Los Angeles Times*, 17 December 1950.

25. "Report on General Panel Corporation Prefabricated House," from Los Angeles Corps of Engineers, U.S. Army, to Division Engineer, South Pacific Division, Oakland, Calif., 14 December 1948, reviews the house from every technical point of view and reports "The quality of the manufactured product is excellent." Adjustment for a 50-lb snow load necessitated a specially designed roof truss. There was room in the panel to increase the fiberglass insulation threefold thereby giving adequate thermal protection.

26. Wachsmann, in his autobiography, recounts the following: "Even after I had left the Company, I was sent to the Atomic Energy Comission site in Los Alamos. They needed 3,000 houses at once. They could only issue a letter of intent if the Company was able to produce a bank credit of about 5 million dollars. But the bank said since this was a very unorthodox case, they should have a letter of intent first, it was a real vicious circle which never could be resolved."

27. Letter Federal Housing Agency/General Panel Corporation, California, 6 May 1959.

28. An exact figure of the total investment is not available. Mendelsohn (interview 16 May 1981) indicated that the RFC loan was for $4 million, and that private investment of 10 percent of that sum, namely $400,000, was required. If one adds to this the mortgage on the property, then we get closer to the figure of $6 million, which is Wachsmann's estimate (in his draft autobiography). Wohlstetter (letter to author, 23 June 1981) considered that future investment demanded not just "a few million," but something in the order of magnitude of "hundreds of millions."

29. Dun and Bradstreet report, 16 December 1948.

30. This information was conveyed to me verbally by Reginald Isaacs. Wachsmann himself, in his autobiography, says that he left the company "without regret."

31. See "Lustron Homes," a paper prepared at the University of California, Berkeley, by Floyd E. Barwig, and printed in Bender, *A Crack in the Rear-View Mirror*, 1973, pp. 58–61.

32. White, in *Prefabrication: A History*, 1965, p.v, cites Ralph Waldo Emerson: "All history becomes subjective; in other words, there is properly no History—only Biography."

33. Wachsmann, *The Turning Point of Building*, 1961, p. 9.

34. These projects include the tubular steel Mobilar structure of 1944–45 (initially conceived during the war years in France and reputedly inspired by steel light standards seen at Lyons); the tetrahedral space structure for the U.S. Air Force, 1959; and a series of team studies including a building system based on aluminium, paper, and plastics at the Institute of Design, Chicago, 1953, and a building system based on industrially produced standardized panels, stressing cables, and anchorages at the International Summer Academy, Salzburg, 1958. (All these projects in Wachsmann, *The Turning Point of Building*, 1961.)

35. White, *Prefabrication: A History*, 1965, pp. 300–301.

36. National Homes, of Lafayette, Ind., was established in July 1940. By 1942 it was operating from a 14,500-sq-ft plant and producing 10 houses a day of timber construction with plywood sheathing. Aiming at economy, low cost, and customer satisfaction, National Homes grew into a major organization, which is still in operation today. For an account of the origins and early development of this firm, see " . . . and a Representative Prefabricator," *Architectural Forum*, February 1942, pp. 89ff.

37. "Government-sponsored temporary housing," in White, *Prefabrication: A History*, 1965, pp. 238–249.

38. White, *Prefabrication: A History*, 1965, p. 148.

39. Kelly, *The Prefabrication of Houses*, 1951, p. 60.

40. Department of Housing and Urban Development, *Housing System Proposals for Operation Breakthrough*, 1971.

41. Joseph Carreiro et. al. *The New Building Block: A Report on the Factory-Produced Dwelling Module*, 1968.

42. Walter Gropius, *Apollo in the Democracy: The cultural Obligation of the Architect*, 1968, p. 97.

43. Konrad Wachsmann, "Inside the Bauhaus: An Introduction," manuscript, 1975.

44. Arthur Bernhardt et. al., *Building Tomorrow: The Mobile/Manufactured Housing Industry*, 1980, p. 134.

45. Bernhardt, *Building Tomorrow*, 1980, p. 503. In the supporting environment Bernhardt includes the supply sector, transportation, and finance; in the regulatory environment are land use controls, taxation, building code regulations, and highway regulations.

46. "In the pursuit of these aims I bloodied my nose re-

peatedly in my attempts, first, to give prefabrication for the mass market an early, architect-controlled start and, second, to put visual education on a much broader and more contemporary basis. . . ." Gropius, *Apollo in the Democracy*, 1978, p. 73.

47. Gropius, letter to a group of students, 1964.

Bibliography

Anonymous Articles (in chronological order)

"Neuzeitliche Blockhaus-Bauweise," *Der Holzbau* (supplement to *Deutsche Bauzeitung),* 54:6, 1920, pp. 21–23.

"Die Holzhaus-Siedlungen des 'Wohnungs-Verbandes Gross-Berlin,' " *Der Holzbau,* 54:13, 1920, pp. 49–52.

"Bemerkenswerte Holzbauten von Christoph & Unmack in Niesky in der Ober-Lausitz," *Der Holzbau,* 54:19, 1920, pp. 75–76.

"Probe-Montage eines Holzhauses in der Montage-Halle von Christoph & Unmack," *Der Holzbau,* 55:11, 1921, p. 42.

"Prufung der von der Christoph & Unmack A.-G. Niesky O.-L. im Auftrag des Holzbau-Industriellen-Verbandes errichten Wände (Block-und Tafelwand) auf Wärmeschutz," *Der Holzbau,* 55:4, 1921, pp. 13–16.

"Holzhäuser der 'Holzbau A.G.' in Neuss am Rhein," *Der Holzbau,* 56:2, 1922, pp. 5–8.

"Deutsche und Österreichische Stahlhäuser," *Moderne Bauformen,* July 1927, pp. 5–11.

"Fabrikmässiger Wohnungsbau," *Berliner Lokal-Anzeiger,* 18 March 1928.

"Typisierung des Wohnungsbaues," *Berliner Tageblatt,* 19 March 1928.

"Steel Houses in Germany," *Iron Age,* 5 April 1928, p. 934.

"Mitteilungen aus der Fachwelt," *Moderne Bauformen,* 1929, pp. 72–78.

"Holzhausbau," *Wasmuth's Monatshefte Baukunst & Städtebau,* 14:1930, pp. 564–565.

"Germany Now Has Copper Houses Made in Sections Easily Set Up," *Copper and Brass Bulletin*, 1931, pp. 2–3.

"Portable Steel and Copper House Developed in Germany," *Iron Age*, 12 February 1931, p. 548.

"60 billige zeitgemässe Eigenhäuser," *Bauwelt*, 9, 26 February 1931, pp. 256ff, pl. 1–60.

" 'Die Wohnung Unserer Zeit' auf der Deutschen Bauaustellung Berlin 1931," *Moderne Bauformen*, July 1931, pp. 329–330, + plates.

"Konrad Wachsmann, Berlin: Holzhausbau," *Moderne Bauformen*, 8:3, 1931, pp. 401–403, 418–419.

"Häuser zu Festen Preise," *Bauwelt*, 9, 3 March 1932, pp. 221–254.

"Hirsch Kupfer. Reorganisation unter Auslandfuhrung," *Berlin Tageblatt*, 199, 27 April, 1932.

"Haus für alle?" *Bauwelt*, 22, 2 June 1932, pp. 539–540.

"Technische Einzelheiten auf der Berliner Sommerschau," *Bauwelt*, 24, 16 June 1932, pp. 590–591.

"Ein Baufachman geht durch die Sommerschau," *Bauwelt*, 26, 30 June 1932, pp. 635–636.

"Das Wachsende Haus," *Bauwelt*, 28, 14 July 1932, p. 688.

"Nicht Mehr Kupferhäuser," *Bauwelt*, 30, 28 July 1932, p. 744.

"Portable Houses of Copper," *Architectural Forum*, 58, February 1933, pp. 145–147.

"Steel Houses," *Architectural Forum*, April 1933, pp. 327–331.

"The Modern Houses of the Century of Progress Exposition," *Architectural Forum*, 59, July 1933, pp. 51ff.

"Technical News and Research: New Housing Designs and Construction Systems," *Architectural Record*, 75:1, January 1934, pp. 12–33.

"The Gropius Exhibition," *The Architect and Building News*, 18 May 1934, p. 181.

"Neuzeitlicher Holzhausbau," *Frankfurter Zeitung*, 9 May 1935.

"Sectional House Developed in Germany," *Architectural Record*, 80, July 1936, p. 70.

"Fresh Approach to Housing: Steel Prefabrication, Flexible Size and Igloo-Like Design; Martin Wagner's House," *Architectural Forum*, 74, February 1941, pp. 87–90.

"Building for Defense," *Architectural Forum*, 75, August 1941, p. 107.

"Houses for Defense . . . Defense Houses at New Kensington, Pa. Gropius and Breuer, Associated Architects," *Architectural Forum*, 75, October 1941, pp. 218–220.

"Houses for Defense," *Architectural Forum*, 75, November 1941, pp. 321–326.

"Prefabrication Gets Its Chance," *Architectural Forum*, 76, February 1942, pp. 81–88.

"Recreational Center Proposed for Key West, Florida. Walter Gropius, Architect, Konrad Wachsmann, Associated," *Architectural Forum*, 77, August 1942, pp. 83–85.

"Prefabrication Up-to-Date," *American Builder*, 65, January 1943, pp. 44–45, 77–79.

"Prefabricated Houses," *Architectural Forum*, 77, June 1943, pp. 89–96.

"A List of Prefabricators," *Architectural Record*, 93, June 1943, pp. 75–79.

"The Month in Building . . . News," *Architectural Forum*, 79, September 1943, pp. 41–42.

"Prefabrication: Modulok," *The Architectural Forum*, 79, September 1943, pp. 65–68.

"Variety of Houses from Identical Prefabricated Units of General Panel Corp.: Designed by Harvard Students," *Pencil Points*, 24, December 1943, pp. 76–77.

"Expansible Prefab House for Post-War. Walter Gropius, Architect," *Architectural Record*, 96, December 1944, p. 69.

"General Panel Corp." *Techniques et architecture*, 5, May–June 1945, pp. 134–135.

"Department Store Subdivision Contains Eight Preassembled Houses," *Architectural Forum*, 84, February 1946, pp. 7–8.

"Prefabrication," *Architectural Forum*, 84, April 1946, pp. 137–142, 190.

"Review of Current Techniques," *Prefabricated Homes*, November–December 1946, pp. 26–27.

"Newcomers: General Panel Is First to Benefit from Everything Government Has to Give," *Architectural Forum*, 86, January 1947, p. 16.

"The Industrialized House: General Panel Corporation," *Architectural Forum*, 86, February 1947, pp. 115–120.

"House in 'Industry': A System for the Manufacture of Industrialized Building Elements by Konrad Wachsmann and Walter Gropius," *Arts and Architecture*, 64, November 1947, pp. 28ff.

"This Factory Built House Came True," *Los Angeles Times*, 17 December 1950.

"Haavara," *Encyclopaedia Judaica*, Jerusalem, Keter Publications, 1972, v. 7.

"Hirsch, Aron Siegmund (1858–1941?)," *Encyclopaedia Judaica*, Jerusalem, Keter Publications, 1972, v. 8.

"Metals and Mining," *Encyclopaedia Judaica*, Jerusalem, Keter Publications, 1972. v. 11.

Books and Articles

Abel, Alan O., et al. *The Industrialized House: A Study of Its Development and Prospects*. Harvard University, Graduate School of Business Administration, Cambridge, 1952.

Anderson, David Christian. *Architecture as a Means for Social Change in Germany, 1918–33*. Ph.D. diss., University of Minnesota, Minneapolis, 1972.

Anderson, Stanford Owen. *Peter Behrens and the New Architecture of Germany, 1900–1917*. Ph.D. diss., Columbia University, New York, 1968 (copyright 1970).

Bauer, Catherine. *Modern Housing*. Boston: Houghton Mifflin, 1934.

Bemis, Albert Farwell. *The Evolving House*. V. 3: *Rational Design*. Cambridge, Mass.: Technology Press, 1936.

Bender, Richard. *A Crack in the Rear-View Mirror*. New York: Van Nostrand Reinhold Company, 1973.

Benson, A. C., Jones, F. M., Vaughan, J. E. "Early Example of Prefabrication." *Royal Institute of British Architects Journal*, June 1963, p. 251.

Benton, Tim, and Charlotte (with Dennis Sharp). *Form and Function: A Source Book for the History of Architecture and Design, 1890–1939*. London: Crosby, Lockwood, Staples, c. 1975.

Bernhardt, Arthur D., et al. *Building Tomorrow: The Mobile/Manufactured Housing Industry*. Cambridge: The MIT Press, 1980.

Blecken, Heinrich. "Stahlhäuser." *Konstruktion und Ausführung (Deutsche Bauzeitung)*, 20, October 1926, pp. 149–151.

Blecken, Heinrich. "Neuzeitlicher Stahlhausbau," *Moderner Wohnbau (Deutsche Bauzeitung)*, 3, March 1929, pp. 31–35.

Block, F. *Probleme des Bauens*. V. 1. Potsdam, Muller & Kiepenheuer, 1928.

Bockman, Walter. "Die veranderbare Wohnung." *Bauwelt*, 12, 19 March 1931, pp. 394–395.

Boone, Andrew R. "Look, No Nails: House Hooks Together." *Popular Science*, November 1950, pp. 138–139.

Borngräber, Christian. "The Socialist Impact of the New Architecture in Germany." *Architectural Association Quarterly*, 11:1, 1979, pp. 39–43.

Borngräber, Christian. "Foreign Architects in the USSR." *Architectural Association Quarterly*, 11:1, 1979, pp. 50–62.

Briggs, Ella. "Praktische Fragen zur Erwerbslosensiedlung." *Bauwelt*, 44, 29 October 1931, pp. 1394–1396.

Bruce, Alfred, and Sandbank, Harold. *A History of Prefabrication*. Raritan, N.J.: John B. Pierce Foundation, Housing Research Division, 1945.

Buddensieg, Tilmann, and Rogge, Henning. *Industriekultur: Peter Behrens und die AEG*. Berlin: Gebr. Mann Verlag, 1978.

Bullock, Nicholas. "Housing in Frankfurt 1925 to 1931 and the New Wohnkultur," *Architectural Review*, 163:976, June 1978, pp. 335–342.

Burchard, John. "Research Findings of Bemis Industries, Inc." *The Architectural Record*, 75:1, January 1934, pp. 3–8.

Burchard, John. "Survey of Efforts to Modernize Housing Structure." In Bemis, A. F., *The Evolving House*. V. 3. Cambridge, Mass: Technology Press 1936.

Burton. *Cape Colony for the Settler*. 1903.

Carreiro, Joseph et al. *The New Building Block: A Report on the Factory-Produced Dwelling Module*. Research Report No. 8. Center for Housing and Environmental Studies, Cornell University, Ithaca, 1968.

Christoph und Unmack. *Nordischer Blockhäuser*. Catalog XVII. Niesky: Christoph und Unmack, A.G., c. 1926.

Christoph und Unmack. *Schulpavilions, System Doecker*. Catalog H161. Niesky: Christoph und Unmack, A.G., 1933.

Christoph und Unmack. *Krankenpavilions in Holzbauweise. Original-System Doecker*. Catalog H150. Niesky: Christoph und Unmack, A.G., 1934.

Creighton, T. H. *Building for Modern Man*. Princeton University Press, 1949.

Darlington, Ida. "Thompson Fecit." *Architectural Review*, September 1958, pp. 187–188.

Department of Commerce. *Construction of Steel Houses Increasing in Western Germany*. Special Circular No. 703. Iron and Steel Division. Washington, 23 May 1928.

Department of Housing and Urban Development. *Housing Systems Proposals for Operation Breakthrough*. Washington, D.C.: U.S. Government Printing Office, 1971.

Deutsche Kupferhaus Gesellschaft. *All Kupferhaus: Das ideale Einfamilienhaus*. Catalog. Berlin: Deutsche Kupferhaus Gesellschaft, M.b.H., c. 1932–33.

Deutsche Kupferhaus Gesellschaft. *Warum Kupferhäuser für Palästina?* Catalog. Berlin: Deutsche Kupferhaus Gesellschaft M.b.H., 1933.

Deutscher Werkbund. *Bau und Wohnung*. Stuttgart: Wedekind, 1927.

Deutsche Werkstätten. *De-We Holzhäuser*. Catalog. Hellerau: Deutsche Werkstätten, A.G., November 1929.

Diamant, R. M. E., *Industrialized Building 2*. London: Iliffe Books Ltd., 1965.

Dietz, A. G. H., and Cutler, L. S., eds. *Industrialized Building Systems for Housing*. Cambridge: The MIT Press, 1971.

Eichler, Ned. *The Merchant Builders*. Cambridge: The MIT Press, 1982.

Eisler, Max. "Das 'Wachsende Haus' in Wien." *Moderne Bauformen*, 1932, pp. 289–308.

Eliot, T. S. *Selected Poems*. Middlesex: Penguin Books, 1948.

Emmet, B. *Catalogues and Counters*. Chicago: University of Chicago Press, 1950.

Export Bau-und Handels-Gesellschaft. *Ein Haus für Sie in Palästina*. Böhler Catalog. Berlin, 1934.

Fairbairn, W. *On the Application of Cast and Wrought Iron to Building Purposes*, n.d.

Feilchenfeld, Werner. *Jewish Trade Policy: On the Basis of Transfer Agreements with Central and East European Countries*. Tel Aviv: Haaretz Press, 1938.

Fitch, James Marston. *Walter Gropius*. New York: Braziller, 1960.

Franciscono, Marcel. *Walter Gropius and the Creation of the Bauhaus in Weimar*, Urbana, University of Chicago Press, 1971.

Friedler, Ya'acov. ''Remembering Wannsee'', *The Jerusalem Post*, 19 January 1982, p. 5.

Fry, E. Maxwell. ''Walter Gropius.'' *Architectural Review*, 117:619, March 1955, pp. 155–157.

Fuller, R. Buckminster. ''Dymaxion Houses.'' *Architectural Record*, 75, January 1934, pp. 9–11.

Gascard, Dirk, and Canthal, P. M., ''Das Kleinst-Eigenheim.'' *Bauwelt*, 42, 15 October 1931, pp. 1324–1325.

General Panel Corporation. *Your General Panel Home*. Brochure. Burbank, Calif., n.d.

Giedion, Siegfried. *Walter Gropius: Work and Teamwork*. London: The Architectural Press, 1954.

Gould, Jeremy. *Modern Houses in Britain, 1919–1939*. Architectural History Monographs. No. 1. Society of Architectural Historians of Great Britain, 1977.

Gropius, Ise, ''History of the Gropius House in Lincoln, Massachusetts.'' Mimeographed. 1977.

Gropius, Walter. ''Programm zur Gründung einer allgemeine Hausbaugesellschaft auf künsterlisch einheitlicher Grundlage, m.b.H.'' Typescript (summary of memorandum), March 1910.
Translations:
1. ''Gropius at Twenty-six.'' *Architectural Review*, July 1961, pp. 49–51.
2. ''Industrialized Building.'' *Constructor*, August 1969, pp. 17–27 (Reprint from *Arch. Review*).
3. Wingler, *Bauhaus*, 1969, pp. 20–21 (excerpts).

Gropius, Walter. *Internationale Architektur*. Munich: Albert Langen Verlag, 1925.

Gropius, Walter. ''Wie bauen wir billigere, bessere, schönere Wohnungen?'' Manuscript, 1926 (revised 1927). Published *Werkland*, Leipzig, 5, 1926, p. 268. (English translation in Benton, *Form and Function*, 1975, pp. 195–196.)

Gropius, Walter. "Der Grosse Baukasten." *Das Neue Frankfurt,* 1:2, 1927–28, pp. 25–30.

Gropius, Walter. "Systematische Vorarbeit für rationellen Wohnungsbau," *Bauhaus,* 2, 1927, pp. 1–2. (English translation in Wingler, *Bauhaus,* 1969, pp. 126–127.)

Gropius, Walter. "Der Architekt als Organisator der modernen Bauwirtschaft und seine Forderungen an die Industrie." 1927. In F. Block. *Probleme des Bauens.* V.1. Potsdam: Muller & Kiepenheuer, 1928.

Gropius, Walter. "Der Gedanke der Rationalisierung in der Bauwirtschaft." *Reichsforschungsgesellschaft für Wirtschaftlichkeit im Bau- und Wohnungswesen,* April 1929, pp. 14–26.

Gropius, Walter. "Siedlung in Stuttgart, Weissenhof." *Reichsforschungsgesellschaft für Wirtschaftlichkeit in Wohnungsbau,* 6, 1929, pp. 104–106, 133, 136, 144, 147, 148 (special number).

Gropius, Walter. "Versuchssiedlung in Dessau." *Reichsforschungsgesellschaft für Wirtschaftlichkeit in Wohnungsbau,* 7, 1929, pp. 1–136 (special number).

Gropius, Walter. *Bauhausbauten Dessau.* Bauhausbücher, 12. Munich: Verlag Albert Langen, 1930.

Gropius, Walter. "The Formal and Technical Problems of Modern Architecture and Planning."

1. *The Architect and Building News,* 18 May 1934, pp. 203–206.

2. *Journal of the Royal Institute of British Architects,* 19 May 1934, pp. 679–694.

Gropius, Walter. "Theaterbau." *Convegno di Lettere, Ottobre 1934.* Rome: Reale Accademia d'Italia, Fondazione Alessandro Volta, 1935, pp. 154–162.

Gropius, Walter. *The New Architecture and the Bauhaus.* Trans. P. Morton Shand. London: Faber & Faber, 1935.

Gropius, Walter. *Rebuilding Our Communities.* Chicago: Paul Theobald, 1945.

Gropius, Walter. "Progress in Housing." *New York Times,* 2 March 1947.

Gropius, Walter. "Prefabrication: A Freedom from Limitations." In T. H. Creighton. *Building for Modern Man.* Princeton: Princeton University Press, 1949, pp. 41–45.

Gropius, Walter. *The Scope of Total Architecture.* New York: Harper, 1955.

Gropius, Walter. *Apollo in the Democracy: The Cultural Obligation of the Architect.* New York: McGraw-Hill, 1968.

Hahn, (Marine-Baurat). "Neuzeitliche Blockhausbauten." *Deutsche Bauzeitung* (supplement *Der Holzbau*), 54:21, 1920, pp. 81–84.

Handlin, Oscar. *Truth in History.* Cambridge: Belknap/Harvard University Press, 1979.

Harrison, D. Dex, and Albery, Whiting. *A Survey of Prefabrication.* London: Ministry of Works, 1945.

Haskell, Douglas. "Prefabrication: Assembly Lines Reach Out for Markets." *The Architectural Record,* 93, June 1943, pp. 62–69.

Heinicke, Erich. "Das 'Wachsende Haus.' " *Deutsche Bauzeitung,* 11 May 1932, pp. 392–398.

Heinicke, Erich. "Zum Thema 'Wachsende Haus.' " *Deutsche Bauzeitung,* 22 June 1932, pp. 503–510.

Herbert, Gilbert. *The Synthetic Vision of Walter Gropius.* Johannesburg: Witwatersrand University Press, 1959.

Herbert, Gilbert. "The Architectural Design Process." *British Journal of Aesthetics,* 6:2, April 1966, pp. 152–171.

Herbert, Gilbert. *Holism, the Ecosystem and Architecture: Towards a Philosophy of Architectural Design.* Durban: University of Natal Annual Discourse Series, 1975.

Herbert, Gilbert. *Pioneers of Prefabrication: The British Contribution in the Nineteenth Century.* Baltimore: Johns Hopkins University Press, 1978.

Herbert, Gilbert. "Israel." In Sanderson, W. (ed.). *International Handbook of Contemporary Developments in Architecture.* Westport, Conn.: Greenwood Press, 1981.

Herbert, Gilbert, and Stark, Paul. *Manning Houses: Their Provenance and Extent in South Australia. A Preliminary Inventory.* Architecture Working Paper No. 4. University of Adelaide, 1979.

Herrey, Herman. "At Last We Have a Prefabrication System Which Enables Architects to Design Any Type of Building with 3-Dimensional Modules." *Pencil Points,* 24, April 1943, pp. 36–47.

Hirsch, Joseph. *The Book of the Late Reb Binyamin Hirsch.* Jerusalem: Tarshish Publishers, 1948 (in Hebrew).

Hirsch Kupfer-und Messingwerke. *All Kupferhaus: Das ideale Einfamilienhaus.* Catalog. Berlin: Hirsch Kupfer-und Messingwerke A.G., c. 1931.

Hirsch Siegmund. *Revolution im Messing: 1908–1928*. Lammersdorf, Otto Junker, 1967.

Hodes, D. A., and Jensen, G. F. *The Case for Industrialized Housing—Re-Examined*. Housing Research Corporation, 1973.

Issacs, Reginald R. *Walter Gropius: Der Mensch und sein Werk*. V. 1. Berlin: Gebr. Mann Verlag, 1983.

Kalitzky, B. "Mivneh Plada b'Shitat Boehler" (Hebrew). *Habinyan Hamisrach Hakarov*, 2:1, 1937, pp. 10–11.

Keith, Nathaniel S. *Politics and the Housing Crisis since 1930*. New York: Universe Books, 1973.

Kelly, Burnham. *The Prefabrication of Houses*. Cambridge, Mass.: Technology Press/Wiley, 1951.

Klein, Alexander. "Die Kleinwohnung als Wirtschaftliches, Wohntechnisches und Raumgestalterisches Problem." *Deutsche Bauzeitung*, 29 April 1931, pp. 213–216.

Kramer, Ferdinand. "Das neue Frankfurt." *Architectural Association Quarterly*, 11:1, 1979, pp. 44–49.

Kress, Fritz. "Die Konstruktionen und Holzbausysteme." *Moderne Bauformen*, 1933, pp. 596–634.

Kuhberg, Leo. "Die Holzbauweise in der technischen Zukunft Deutschlands." *Bauwirtschaft und Baurecht (Deutsche Bauzeitung)*, 24, December 1926, pp. 121–123.

Lane, Barbara Miller. *Architecture and Politics in Germany, 1918–45*. Cambridge: Harvard University Press, 1968.

Le Corbusier. *Towards a New Architecture*. Trans. Frederick Etchell. London: Architectural Press, 1927 (1946 edition).

Lewcock, Ronald. *Early-Nineteenth-Century Architecture in South Africa*. Cape Town: Balkema, 1963.

Llewellyn, F. T., and Speller, F. N. "Steel in Residence Construction." *The Architectural Record*, 73, June 1933, pp. 438–442.

Lion, A. "Die ersten Wohnungsbauten aus Betonplatten in Deutschland." *Konstruktion und Ausfuhrung (Deutsche Bauzeitung)*, 15 August 1926, pp. 112–115.

Lion, A. "Stahl-Wohnhäuser." *Zeitschrift des Österr. Ingenieur-und Architekten-Vereines*, 17–18, 1927, pp. 156–158.

Ludowici, W. ''Die Finanziellen Vorteile des Wachsendes Hauses.'' *Deutsche Bauzeitung*, 20 July 1932, p. 584.

Manning, Fowler, ''Selling the Prefabricated House?—Here Are Some of the Unsolved Problems.'' *American Builder*, 66, February 1944, pp. 51, 96, 98–100.

Marcus, Ernst. ''The German Foreign Office and the Palestine Question in the Period 1933–39.'' *Yad Vashem Studies*, 2, 1958, pp. 179–204.

Mendelsohn, Erich. *Russland, Europa, Amerika: Ein architektonischer Querschnitt*. Berlin: R. Mosse, 1929.

Moholy-Nagy, Sibyl. ''The Diaspora.'' *Journal of the Society of Architectural Historians*, 24:1, March 1965, pp. 24–25.

Moore, Richard. ''An Early System of Large-Panel Building.'' *RIBA Journal*, 76:9, September 1969, pp. 383–386.

Murche, Archt. ''Doppelhaus als 'wachsendes' Kleinsiedlungshaus.'' *Bauwelt*, 44, 29 October 1931, p. 1389.

Naylor, Gillian. *The Arts and Crafts Movement: A Study of Its Sources, Ideals, and Influences on Design Theory*, Cambridge: The MIT Press, 1980.

Neutra, Richard. *Wie Baut Amerika?* Stuttgart: Julius Hoffman, 1927.

Newell-Emmett Co. *Prefabricated Housing: Bibliography and Briefed Articles from Business, Technical and Consumer Publications, 1927–1942*. New York, 1943.

Pehnt, Wolfgang. ''Gropius the Romantic.'' *Art Bulletin*, September 1971, pp. 379–392.

Perret, Louis. ''L'Immeuble Normalisé a Montage Rapide.'' *Arts et metiers*, 94, July 1928, pp. 241–246.

Peterson, Charles E. ''Early American Prefabrication.'' *Gazette des Beaux-Arts*, 6th ser., no. 33, 1948, pp. 37–46.

Peterson, Charles E. ''Prefabs for the Prairies.'' *Journal of the Society of Architectural Historians*, November 1952, pp. 28–29.

Peterson, Charles E. ''Prefabs in the California Gold Rush, 1849.'' *Journal of the Society of Architectural Historians*, December 1965, pp. 318–324.

Peterson, J. L. ''History and Development of Precast Concrete in the United States.'' *Journal of the American Concrete Institute*, 25:6, February 1954, pp. 477–496.

Petzold, Erich. ''Zum Problem des Wachsenden Häusen.'' *Deutsche Bauzeitung*, 17 February 1932, p. 141.

Pevsner, Nikolaus. *Pioneers of Modern Design: from William Morris to Walter Gropius*. Middlesex: Penguin Books, rev. ed., 1960. (First published by Faber and Faber under the title *Pioneers of the Modern Movement*, 1936).

Rasch, H. and B. *Wie bauen?* Stuttgart: Wedekind, c. 1927.

Richards, J. M. (ed.). *Who's Who in Architecture*. London: Weidenfeld and Nicholson, 1977.

Riemerschmid, Richard. ''Holzhäuser.'' *Bauwelt*, 40, October 1932, pl. 5–8.

Robertson, Howard. ''The First Nazi Exhibition, Berlin.'' *The Architect and Building News*, 18 May 1934, pp. 193–195.

Roth, Cecil. *History of the Jews*. New York: Schocken Books, 1961.

Schadewald, M. ''Wachsende Haus-Sinkende Schuld.'' *Bauwelt*, 41, 8 October 1931, pp. 1291–1293.

Schmid, Alfred. ''Wohnhäuser nach dem Stahlwandsystem.'' *Zeitschrift des Österr. Ingenieur-und Architekten-Vereines*, 19–20, 1927, pp. 174–176.

Schmidt, Otto. ''Verstellbare Wande für variable Wohnung.'' *Bauwelt*, 36, 3 September 1931, pp. 1156–1157.

Sharp, Dennis. *Sources of Modern Architecture*. London: Lund Humphries (for Arch. Assoc.) 1967.

Shiffman, J. ''The Building Industry in Palestine.'' *Palestine and Middle East Economic Magazine*, 7–8, 1933, p. 288.

Siedler, Jobst. ''Neues Bauen,'' and Other Technical Notes. *Bauwelt*, 1931, nos. 26, 28–36, 38, 39, 42, 45–48, 50, 51.

Sommerfeld, Adolf. ''Was will der 'Deutsche Holzbau-verein'?'' *Der Holzbau*, 54:1, 1920, p. 1.

Sonn, Edith. ''Factory-Cut But Flexible.'' *New York Times Magazine*, 11 August 1946, p. 34.

Spiegel, Hans. *Das Stahlhausbau*. 2 vols. Leipzig: Alwin Frohlich Verlag, 1928, 1929.

Spiegel, Hans. ''Stahlbauweisen für das kleine Einzelwohnhaus.'' *Deutsche Bauzeitung*, 6 July 1932, pp. 553–560.

Stowell, Kenneth Kingsley. "Housing the Other Half." *Architectural Forum*, 56, March 1932, pp. 217–220.

Stratmann, Mechtild. "Housing Policies in the Weimar Republic." *Architectural Association Quarterly*, 11:1, 1979, pp. 16–23.

Sturges, W. Knight. *Introduction: The Origins of Cast Iron Architecture in America*. New York: Da Capo Press, 1970.

Taut, Bruno. "Grenzen der Wohnungsverkleinerung." *Deutsche Bauzeitung*, 29 April 1931, pp. 210–212.

Thomson, David, *Europe since Napoleon*. London: Pelican Books, 1966.

U. S. Information agency. *Heritage of Freedom: A Brief History of the United States*. V.3, pt.2. Washington, D.C., n.d.

Vaux, F. Wilson. *Tomorrow's Homes*, Trenton, N.J.: Homasote Co., 1939.

Von Moltke, Willo. "Prefabricated Panels for Packaged Buildings." *Architectural Record*, 93, April 1943, pp. 50–53.

Wachsmann, Konrad. *Holzhausbau*. Berlin: Wasmuth, 1930.

Wachsmann, Konrad. "Kleiner und Grosse Bauten in neuer Holzbautechnik." *Bauwelt*, 50, 10 December 1931, pp. 1559–1574.

Wachsmann, Konrad. "Machine Energy: The Technique of Our Time." In T. H. Creighton. *Building for Modern Man*. Princeton: Princeton University Press, 1949, pp. 46–48.

Wachsmann, Konrad. "Zum 75. Geburtstag von Walter Gropius." *Baukunst und Werkform*, 11:5, 1958, pp. 284–286.

Wachsmann, Konrad. *The Turning Point of Building*. New York: Reinhold, 1961.

Wachsmann, Konrad. " 'Inside the Bauhaus': An Introduction." Unpublished manuscript, 1975.

Wagner, Martin. *Das Wachsende Haus*. Berlin: Deutsches Verlagshaus Bong & Co., 1932.

Wagner, Martin. *Prefabricated Housing*. Unpublished manuscript (typescript amended by hand for publication), 1939–40, foreword dated 6 July 1941. Drawings (missing from ms) ascribed to Bernard Wagner and Heinrich Shapiro. Trans. assisted by Oscar Sutermeister.

Ward, Robertson, "Konrad Wachsmann: Towards Industrialization of Buildings." *AIA Journal*, 57:3, March 1972, pp. 33–42.

Ward, Robertson, Cruz, Stephen, and Wang, Elizabeth. *The Complete Project Works of Konrad Wachsmann, May 16, 1901–November 25, 1980.* Draft index for microfilm collection, Huntington Library, in association with the Smithsonian Museum.

White, R. B. *Prefabrication: A History of Its Development in Great Britain.* London: Her Majesty's Stationery Office, 1965.

Whittick, Arnold. "Le Corbusier." In Wolfgang Pehnt (ed.). *Encyclopaedia of Modern Architecture.* London: Thames and Hudson, 1963.

Wiedenhoft, Ronald Victor. *Workers' Housing in Berlin in the 1920s: A Contribution to the History of Modern Architecture.* Ph.D. diss. Columbia University, New York, 1971.

Wilhelm, Karin. "From the Fantastic to Fantasy." *Architectural Association Quarterly,* 11:1, 1979, pp. 4–15.

Wilson, F. Vaux. "New Approach to 'Packaged' House Recognizes Architects." *Architectural Record,* 82, August 1931, pp. 86–87.

Winchell, Prentice. "The Dwelling of Tomorrow: An Economic Study in Residential Construction Showing Why a Growing Use of Iron and Steel is Inevitable." *Iron Age,* 117, 1927, pp. 613–615, 686–687, 766–768, 840–841, 930–931, 992–993.

Windsor, Alan. *Peter Behrens.* London: Architectural Press, 1981.

Wingler, Hans. *Bauhaus.* Cambridge: The MIT Press, 1969.

Wurm, Heinrich. "Vorgefertige Bauwerke des 19. Jahrhunderts: Der Stand der Forschung." *Technikgeschichte Bd.,* 33:3, 1966, pp. 228–255.

Wurm, Heinrich. "Die Industrialisierung des Holzhausbaues: Christoph und Unmack." *Tradition: Zeitschrift für Firmengeschichte,* 14, 1969, pp. 198–211.

Wurster, Catherine Bauer. "The Social Front of Modern Architecture in the 1930s." *Journal of the Society of Architectural Historians,* 24:1, 1965, pp. 48–52.

Yisraeli, David. "The Third Reich and the Transfer Agreement." *Journal of Contemporary History,* 2, 1971, pp. 129–148.

Young, C. D. & CO. *Illustrations of Iron Structures for Home and Abroad, Consisting of Stores, Dwelling-Houses, Markets, Arcades, Railway Stations, and Roofing, &c. &c., Constructed of Wrought and Cast Iron and Corrugated Sheets, Manufactured by Charles D. Young & Co.* Edinburgh, c. 1856.

Illustration Credits

Illustrations by courtesy of:

American Institute of Architects Journal, 9.7 (March 1972)

Architectural Forum, 7.7 (February 1941), 9.12 (February 1947), 9.11 (February 1942)

Architectural Record, 7.3 (January 1934)

Bauhaus-archiv, Berlin, frontis, 2.6 top, 6.1

Bauwelt, 3.6 (17 November 1932), 3.7 (3 March 1932), 3.9 (October 1932), 6.3 (26 February 1931)

Bernhardt, *Building Tomorrow,* (MIT Press), 10.4 (photos by Schult Mobile Homes), 10.5

Block, *Probleme des Bauens* (Muller & Kiepenheuer), 2.3, 2.5

Böhler, *Massivebauten für Palästina,* catalog, 5.7

Bruce and Sandbank, *A History of Prefabrication* (J. B. Pierce Foundation), 7.2, 7.4 (photo by Kaufman-Sabry), 7.5, 7.6

Busch-Reisinger Museum, Harvard University, 3.20, 4.9–4.13, 9.2, 9.3

Christoph und Unmack, *Nordischer Blockhäuser,* catalog, 3.13

Deutsche Bauzeitung, 2.4 (15 August 1926), 3.5 (6 July 1932), 4.3 (15 July 1931)

Deutsche Kupferhaus, *All Kupferhaus,* catalog, 4.19

Deutsche Kupferhaus, *Warum Kupferhäuser für Palästina?* catalog, 5.1

Deutscher Werkbund, *Bau und Wohnung* (Wedekind), 2.8

Deutsche Werkstätten, *De We*, catalog, 3.9

Diamant, *Industrialized Building*, (Iliffe Press), 10.3 (Butterworths)

Donchin, Mark, 10.1

Eitan, Dov, 5.9

Fitch, *Walter Gropius* (Braziller), 4.15

General Panel, *Your General Panel Home*, brochure, 9.17

Gropius, *Rebuilding Our Communities*, (Theobald), 8.8, 9.5

Gropius, *Apollo in the Democracy* (McGraw-Hill), 2.1, 2.2 (Gropius family)

Herbert, Gilbert, 1.1 (after Loudon)

Hirsch Kupfer, *All Kupferhaus*, catalog, 4.4, 4.5, 4.7, 4.8

Iron Age, 4.2 (12 February 1931)

Judische Rundschau, 5.4

Katz, Walter, 5.10

Loeb, Mrs. Max, 5.11

Moderne Bauformen, 3.14, 3.18

Naumburg, Menahem, 5.6

Pencil Points. 8.7, 9.1 (April 1943, photos by General Panel Corp.), 9.4 (December 1943, Reinhold Publishing)

RIBA Journal, 1.2 (September 1969)

Scientific American, 3.12 (June 1886)

Spiegel, *Stahlhausbau* (Alwin Frohlich Verlag), 1.3 top, 2.6 bottom, 2.7, 3.1–3.3, 7.1

Technion: Israel Institute of Technology (Faculty of Architecture and Town Planning), 4.6, 5.3, 5.5, 5.8 (drawing by Oded Lilienthal)

Techniques et architecture, 9.6 (May–June 1945)

Tuchler family, 5.2

Wachsmann Archives (Mrs. Judith Wachsmann), 8.1, 8.2, 8.3 (photo by Heinz), 8.4–8.6, 8.10, 9.9 (photo by Karsten), 9.10 (photo by Whittington), 9.13, 9.14 (photo by Meriman), 9.15, 9.16, 9.18

Wachsmann, *Holzhausbau* (Wasmuth), 3.10, 3.15–3.17, 3.19

Wagner, *Das Wachsende Haus* (Bong & Co.), 4.14, 4.16–4.18

White, *Prefabrication: A History* (HMSO), 1.3 bottom, 1.4 (Building Research Establishment, U.K. Crown copyright), 10.2 (photo by Aluminium Federation)

Zeitschrift des Österr. Ingenieur- und Architekten Vereines, 3.4 (17–18, 1927)

Index